Influential Passengers

Influential Passengers

Inherited Microorganisms and Arthropod Reproduction

Edited by

Scott L. O'Neill

Department of Epidemiology and Public Health,
Yale University School of Medicine, USA

Ary A. Hoffmann

School of Genetics and Human Variation,
La Trobe University, Australia

and

John H. Werren

Department of Biology, University of Rochester, USA

OXFORD NEW YORK TOKYO
OXFORD UNIVERSITY PRESS
1997

Oxford University Press, Great Clarendon Street, Oxford OX2 6DP

Oxford New York
Athens Auckland Bangkok Bogota Bombay
Buenos Aires Calcutta Cape Town Dar es Salaam
Delhi Florence Hong Kong Istanbul Karachi
Kuala Lumpur Madras Madrid Melbourne
Mexico City Nairobi Paris Singapore
Taipei Tokyo Toronto Warsaw
and associated companies in
Berlin Ibadan

Oxford is a trade mark of Oxford University Press

Published in the United States
by Oxford University Press Inc., New York

A catalogue record for this book is available from the British Library

Library of Congress Cataloging in Publication Data
(Data available)
ISBN 0 19 857786 9 (Hbk)
0 19 850173 0 (Pbk)

Typeset by Downdell, Oxford
Printed in Great Britain by
Biddles Ltd, Guildford & King's Lynn

Preface

Symbiotic relationships between microorganisms and arthropods are extremely common. However, with some exceptions the study of these interactions has been largely neglected. This can be attributed to the inherent technical difficulty of working with these organisms and the lack of a traditional discipline within the biological sciences which specifically addresses this subject. As a result the study of these systems has remained on the fringes of genetics, microbiology, entomology, invertebrate biology, and evolutionary biology. In recent years technical advances have made many of these systems more accessible and there is a renewed interest in studying them. This rapid increase of symbiosis research is particularly apparent in the study of inherited microorganisms which alter reproduction in infected hosts. However the interdisciplinary nature of this research makes for a fragmented literature which can at times be confusing. In this book we have brought together and synthesized into one volume much of the research on reproductive symbionts. We have provided a theoretical evolutionary framework with which to consider these systems, as well as an applied perspective on how these systems might be used in pest control.

We are very grateful for the work of each of the contributors to this volume as well as to Keith Binnington who initiated the project, François Rousset and Konstantinos Bourtzis for reviewing proofs, and all of our colleagues who have helped us minimize the inevitable errors associated with such an undertaking. We would like to dedicate this book to Marshall Hertig, Hannes Laven, Janice Yen, and Ralph Barr, pioneers in the study of reproductive parasites. These four individuals used diverse approaches to examine reproductive symbionts and laid the foundation for understanding *Wolbachia* mediated cytoplasmic incompatibility.

May 1997

Scott O'Neill
Ary Hoffmann
Jack Werren

Contents

Contributors

Chris F. Curtis Department of Medical Parasitology, London School of Hygiene and Tropical Medicine, London WCIE 7HT, UK.

Ary A. Hoffmann School of Genetics and Human Variation, La Trobe University, Bundoora, VIC 3083, Australia.

Gregory D. D. Hurst Department of Genetics, University of Cambridge, Downing Street, Cambridge CB2 3EH, UK.

Laurence D. Hurst School of Biology and Biochemistry, University of Bath, Claverton Down, Bath BA2 7AY, UK.

Michael E. N. Majerus Department of Genetics, University of Cambridge, Downing Street, Cambridge CB2 3EH, UK.

Scott L. O'Neill Department of Epidemiology and Public Health, Yale University School of Medicine, New Haven, CT 06520, USA.

Thierry Rigaud Université de Poitiers, Laboratoire de Biologie Animale, UMR CNRS 6556, 40 Avenue du Recteur Pineau, F-86022 Poitiers Cedex, France.

Steven P. Sinkins Department of Epidemiology and Public Health, Yale University School of Medicine, New Haven, CT 06520, USA.

Richard Stouthamer Department of Entomology, Wageningen Agricultural University, PO Box 8031, 6700EH Wageningen, The Netherlands.

Michael Turelli Section of Evolution and Ecology and Center for Population Biology, University of California Davis, CA 95616, USA.

John H. Werren Department of Biology, University of Rochester, Rochester, NY 14627, USA.

1 The evolution of heritable symbionts

John H. Werren and Scott L. O'Neill

1.1 Introduction

Symbiotic microorganisms are extremely widespread in nature, having intimate and often obligatory associations with their 'host' species. Despite the near ubiquity of arthropod symbionts, their study has been constrained by their fastidious nature and inability to be cultured out of the arthropod host. Recent advances in molecular biology have provided new tools for symbiosis research and stimulated new investigations of many symbiont systems that were described many years ago but were previously difficult to investigate in detail.

The term 'symbiosis' in its most general (and original) sense refers to the intimate 'living together' of dissimilar organisms (de Bary 1879). Symbiont interactions with hosts have been traditionally classified as mutualistic (beneficial), parasitic (harmful) or commensal (neutral). However, assigning a symbiotic association to these groupings is often problematic. The associations between symbiont and host are complex and can shift between the different states both over time and depending upon the particular phenotype being considered (Clay 1988; Saffo 1991). Furthermore, it is not entirely clear to what extent mutualistic relationships are reciprocal. Although the benefits to the host are often clear and can be established by experimental elimination of the symbiont, it is much less clear whether mutualistic symbionts truly benefit from the association. It has been argued that such symbionts may better be considered slaves of the host than mutualists (Maynard Smith and Szathmary 1995), although domesticated microorganisms may be a better analogy. Regardless of the relationship between symbiont and host (whether mutualistic, parasitic, or exploitative), there is considerable scope for reciprocal manipulation by both parties.

Symbionts are transmitted to new hosts by many different mechanisms, ranging from completely horizontal (infectious) transmission to completely vertical (heritable) transmission. Strictly 'heritable' symbionts are those that have evolved such intimate associations with the host that the dominant mode of transmission is through host reproduction (vertical transmission). Heritable microbes are widespread among animals, being typically intracellular and transmitted in the cytoplasm of eggs, but other mechanisms of transmission also exist (Buchner 1965). Heritable microorganisms have taken symbiosis nearly to its extreme; their survival and reproduction is completely dependent upon the survival and reproduction of the host. For this reason, there is a general view that once a microorganism has strict or nearly strict vertical transmission, it will inevitably evolve to a beneficial (mutualistic) association with its host (Fine 1975; Ewald 1987; Lipsitch *et al.* 1995). However, this widespread and pervasive view neglects an important evolutionary alternative for heritable microorganisms, that of 'reproductive parasitism'. These reproductive parasites, which include cytoplasmic incompatibility, parthenogenesis-inducing, feminizing and some male-killing microorganisms, have evolved to manipulate the reproduction of hosts in ways that enhance transmission of the microorganism, even though they can be detrimental to host reproductive success.

In this chapter we consider the diversity, origins, phylogenetic patterns and evolutionary consequences of heritable symbionts. We review some basic concepts relating to heritable symbionts, focusing on their potential strategies for increase in host populations. An important theme in this chapter is that interactions between heritable symbionts and their hosts are not simply either mutualistic or parasitic, but can be a complex interplay of both. The co-evolution of both conflict and co-operation between symbiont and host has important implications, and may provide a 'motor' for evolution of both parties.

1.2 Heritable symbiont strategies

The long-standing view that microorganisms with exclusive vertical transmission will inevitably evolve to beneficial interactions with the host seems to make intuitive sense (Fine 1975; Ewald 1987; Lipsitch *et al.* 1995). The general conclusion is as follows: because transmission of the symbiont is completely dependent upon survival and reproduction of the host, selection will favour those symbionts that increase host survival and reproduction. Indeed, symbionts that reduce host survival and reproduction cannot be sustained in host populations when they are only vertically transmitted. Although the theoretical treatments are correct given the assumptions made, they have been structured in such a way as to neglect alternative routes for increase of heritable symbionts. Thus, the widespread view that heritable symbionts must be mutualists is incorrect.

1.2.1 Basic model

Consider a vertically transmitted microorganism in a host population. As with most heritable symbionts, we will assume that vertical transmission occurs via the cytoplasm of the egg and that paternal transmission is rare or absent. In addition, horizontal (infectious) transmission does not occur. As will be shown below, for such vertically transmitted symbionts there are four primary routes for maintenance in host populations. These are:

(1) increase fitness of the infected hosts;

(2) increase sex ratio (proportion females) of infected hosts;

(3) decrease fitness of uninfected hosts;

(4) decrease sex ratio (proportion females) of uninfected hosts.

In addition, a heritable symbiont can be maintained by a mixture of heritable and infectious transmission. This will be considered briefly later. Each mechanism described above is characterized with respect to its effect upon the hosts. The first mechanism is classic mutualistic symbiosis. The latter three are mechanisms of reproductive parasitism. Reproductive parasitism can increase the frequency of a symbiont in the host population (even though it is often detrimental to the fitness of the host) because the symbiont gains increased transmission by manipulating host reproduction to its advantage. A general model, presented below, illustrates the basic principles. The model is presented not as a detailed analytical treatment of heritable symbiont dynamics, but as a heuristic model that illustrates the selective forces acting upon these symbionts.

Imagine a population composed of individuals with two different types of cytoplasms, infected (i) and uninfected (u). The microorganism is transmitted vertically through females, with proportion (a) receiving the microbe; the remaining proportion ($1 - a$) revert back to uninfected status. Fitness (survival and fecundity) of infected females is W_i and that of uninfecteds is W_u. Similarly, the proportion of female progeny (primary sex ratio) produced by infecteds and uninfecteds is x_i and x_u respectively. Keep in mind that each of these variables can be a function of the frequency of the symbiont in the population (p). Now, the frequency of the infection in the next generation (p) is simply the number of infected females produced divided by the total number of females produced, and is

$$p = pW_i x_i a/((1 - p)W_u x_u + pW_i x_i).$$

The symbiont will increase in frequency in the population so long as

$$W_i x_i a > (1 - p)W_u x_u + pW_i x_i.$$

The term to the right of the '>' symbol is the average number of daughters produced per female in the population. For the symbiont to increase in frequency, the production of infected daughters by infected females must exceed the average production of daughters per female in the population. Note that if the symbiont controls the sex of the infected individual (rather than the sex ratio of its progeny), then the formula would be:

$$W_i x_i a > (1 - p)W_u x_u + pW_i x_i.(ax_i + (1 - a)x_u).$$

The differences in the two formulae occur simply because the probability of a revertant individual (an uninfected offspring of an uninfected mother) becoming female is different when the symbiont determines the sex ratio of the parent versus the sex of the progeny.

Consider for a moment the conditions for increase of a rare heritable symbiont in a population of uninfecteds, which is the same for both formulae, and is

$$aW_i x_i > W_u x_u.$$

Simply, infected females must produce more infected females than uninfected females produce uninfected females. This formulation also displays the alternative reproductive strategies available to strictly heritable symbionts. Below we will discuss briefly the alternative reproductive strategies. Heritable symbionts are already known to pursue most of the strategies described below. Other strategies discussed are predicted by this basic theory, but have not yet been documented in nature. Again, it is important to keep in mind that mixtures of these strategies can be expected to evolve in vertically transmitted symbionts.

1.2.1.1 Increase host fitness

Heritable symbionts can increase W_i, the fitness of hosts they are in (mutualism). This is the strategy generally thought of for heritable symbionts. Mutualistic symbionts (both heritable and infectious) are widespread among invertebrates (Buchner 1965). Reproductive parasites are also subject to selection to increase host fitness, so long as increasing host fitness does not sufficiently reduce the advantageous fitness effects to the symbiont of manipulating host reproduction or its transmission rate (a). Note that the key parameter here is *fitness of infected females* (not males). Symbionts are not under direct selection to increase the fitness of infected males because males do not transmit the microorganisms.

1.2.1.2 Increase production of infected females

Heritable symbionts can increase the proportion of females produced (x_i). Examples include parthenogenesis-inducing *Wolbachia* in parasitic Hymenoptera (Stouthamer *et al.* 1993; Chapter 4), feminizing *Wolbachia* found in isopods (Rousset *et al.* 1992*a*; Chapter 3) and feminizing microsporidia in shrimp (Dunn *et al.* 1993*a*). Note that the male-killing microbes do not cause an increase in the primary sex ratio (x_i), but rather increase the fitness of infected females by inducing death of their sibling males. Fitness of the infected siblings is presumably enhanced by either freeing resources for them or by a reduction in the level of harmful inbreeding. In some systems, male-killing can provide an inoculum for horizontal transfer of the infection. These scenarios are reviewed in Chapter 5.

1.2.1.3 Decrease fitness of uninfecteds

Heritable symbionts that decrease the fitness of hosts in which they do not occur (W_u) can be selectively favoured. Here we consider three alternative strategies within the general category of reducing the fitness of hosts in which the symbiont does not occur:

(1) cytoplasmic incompatibility phenotype;

(2) killer and harmer phenotype; and

(3) Medea phenotype.

Cytoplasmic incompatibility phenotype

Reducing the fitness of uninfected hosts is precisely the strategy employed by cytoplasmic incompatibility (CI) microorganisms in diploid hosts. So far, all known cases of CI microbes occur in the rickettsial genus, *Wolbachia* (reviewed in Chapter 2). Although the biochemical mechanisms are not known, CI *Wolbachia* within the testes apparently modify developing sperm of infected males. When an egg is fertilized by sperm from an infected male, the same CI-type bacterial strain must be present within the egg to rescue this modification. Otherwise, abnormal mitosis occurs which typically results in zygotic death (in diploid species). Thus, incompatibility can occur between the sperm of infected males and the eggs of uninfected females, or between the sperm of individuals infected with one strain and the eggs of individuals infected with a different strain. In the former case, incompatibility is unidirectional whereas in the latter case it can be either unidirectional or bidirectional. In effect, CI bacteria exploit infected males, which cannot normally transmit the bacterium, to reduce the fitness of uninfected individuals (or individuals infected with different strains).

Turelli (1994) has developed detailed population genetic models for CI bacteria, and Hoffmann and Turelli (Chapter 2) review the evolution and population biology of CI. CI microbes that cause a reduction in the fertility of infected females must exceed a threshold frequency before the advantages of CI induction (reduction in the frequency of uninfecteds) allows their spread through the host population (Caspari and Watson 1959; Turelli 1994). As described by Turelli (1994), even infections with fertility costs (e.g. 1–2 per cent) can readily achieve threshold frequencies by drift, particularly in subdivided populations.

Another interesting possibility is that bacterial strains might arise that do not cause CI, but that are immune to the action of CI bacteria (Turelli 1994). In populations with CI-causing bacteria, these could increase in frequency by parasitizing the action of the CI inducers. That is, they increase in frequency because the CI inducers reduce frequency of uninfecteds, which indirectly benefit both CI inducers and non-inducers that are resistant to CI action. Such bacteria have not yet been found. Although several examples of non-inducing *Wolbachia* are known (Holden *et al.* 1993; Clancy and Hoffman 1996; Giordano *et al.* 1995), these have not been shown to be resistant to the action of resident CI-inducing bacteria.

So far, *Wolbachia* is the only bacterial group known to cause CI. However, in theory any cytoplasmically inherited microorganism might be selected to cause cytoplasmic incompatibility, including mutualistic symbionts that are otherwise beneficial to their hosts. Consider, for example, an obligatory mutualistic endosymbiont such as *Buchnera* in aphids. If a variant symbiont was to arise that also caused CI, this variant could rapidly replace other *Buchnera* strains within host populations.

Killer and harmer phenotypes

Heritable symbionts can be selected to produce transferable products that preferentially kill or harm (e.g. reduce the fitness) of hosts not infected with the symbiont. Such symbionts have been described in Protozoa—the killer endosymbionts found in some *Paramecia* (Preer *et al.* 1974; Pond *et al.* 1989; Heruth *et al.* 1994). These endosymbionts produce a diffusable product (the kappa particle) that, when ingested, kills non-symbiont-bearing *Paramecia.* Symbiont-bearing hosts are immune to this toxin. Because the symbionts reduce the frequency of uninfected *Paramecia* in mixed populations, they will generally increase in frequency relative to the uninfected type, particularly when infected and uninfected hosts compete for limiting resources. Increase of symbiont-bearing hosts will depend upon the fitness costs of harbouring the symbionts relative to the benefits of killing alternative lineages. In addition, symbionts need not cause such extreme effects as killing to be selectively favoured.

Mechanisms that harm uninfected hosts (e.g. reducing growth, survival or fertility) would also suffice and are expected to evolve. However, for killing and harming phenotypes to evolve, there must be a mechanism for delivering the toxic substance to other hosts. For example, *Paramecia* live in a liquid medium, allowing delivery of a toxic substance by diffusion. Possible delivery mechanisms include: (a) diffusion, (b) contact and (c) mating. Based upon the possible delivery mechanisms listed above, we can expect to find killer and harmer symbionts in aquatic arthropods, particularly those breeding in smaller water sources where toxins are more likely to reach targets by diffusion. However, other larval feeding substrates (such as rotting fruit) could also allow diffusion of toxic substances. There is a relatively simple test to determine whether a symbiont has killer or harmer phenotypes; uninfected hosts (or those infected with different symbiont types) will do relatively more poorly when developing in the presence of diffusable substances coming from infected hosts than diffusable substances from uninfected hosts. Mating is a second likely mechanism of toxin delivery. Toxic substances that reduce the fertility of uninfected females could readily be delivered in seminal fluids during mating. Again, a very simple prediction is the reduction in fertility in matings between infected males and uninfected females. Toxin delivery is an alternative mechanism to cytoplasmic incompatibility for reducing the fitness of uninfected females following mating with infected males.

The toxic substance need not be a chemical but could itself be an infectious agent, such as a virus. Such infectious agents could either cause greater harm directly to other hosts or indirectly by harming non-resistant symbionts within other hosts. It remains to be seen whether killing and harming phenotypes are widespread among the symbionts found in arthropods.

Medea phenotype

Medea is a specific phenotype that occurs within the family of infected females, rather than between infected and uninfected families. Consider a symbiont that has incomplete transmission, so that some eggs inherit the symbiont and others do not. Under certain conditions, a symbiont that kills the eggs that did not receive it will be selectively favoured (Fig. 1.1). This scenario is not at all far-fetched. The Medea gene found in *Tribolium* (Wade and Beeman 1994) and spore-killer genes in fungi (Nauta and Hoekstra 1993) function on exactly this principle. Medea (the namesake for this phenotype) was a woman in Greek mythology who killed her own offspring. In *Tribolium*, when the nuclear Medea gene is present in the mother, it must also be present in the zygote or the zygote will die. This effect requires a modification–rescue system, where modification occurs in the parent and rescue must occur within the egg or zygote. It is easy to envision heritable symbionts evolving similar mechanisms. In contrast to nuclear Medea genes, a symbiont Medea will only increase in frequency when killing increases the fitness of the surviving infected progeny, e.g. under sibling resource competition, or when populations are highly subdivided so that the benefits of killing go preferentially to the killing phenotype (Bull *et al.* 1992; Nauta and Hoekstra 1993; Wade and Beeman 1994). The phenotype need not be so extreme as killing. Symbionts that reduce the fitness of uninfected host progeny (but do not necessarily kill them) could also be selectively favoured. The effect could also occur within the reproductive tract of the female (i.e. it need not be after egg-laying). We suspect that it is only a matter of time before symbionts that pursue this strategy are found. Note that a symbiont that also pursues a different strategy (e.g. mutualism, sex-ratio distortion) can be additionally selected to eliminate uninfecteds among the progeny of its host.

1.2.1.4 Decreasing sex ratio of uninfecteds

Heritable symbionts can be selected to reduce the primary sex ratio of hosts in which they do not occur (x_u). This, in fact, is what cytoplasmic incompatibility microbes (*Wolbachia*) do in haplodiploid organisms (Ryan and Saul 1968; Breeuwer and Werren 1990). It is accomplished by causing improper condensation and loss of paternal chromosomes in fertilized eggs, thus causing conversion of diploid (female) zygotes into haploid (male) zygotes (Ryan and

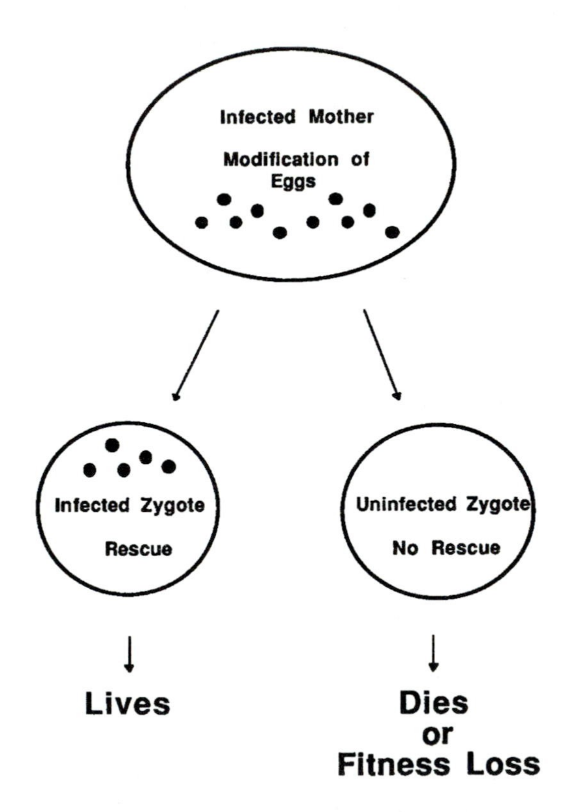

Figure 1.1 Action of Medea symbionts. The Medea effect could evolve in symbionts that have incomplete transmission to progeny. Symbionts present in the mother modify some egg component (e.g. cytoplasm or host genomic contribution) during its development. Symbionts must be present in the developing zygote to rescue this modification, otherwise the zygote dies (or suffers reduced fitness). Nuclear genetic regions are known to induce similar effects (e.g. Medea and meiotic drive chromosomes).

Saul 1968; Reed and Werren 1995). Thus, haplodiploidy provides a ready mechanism for *Wolbachia* to reduce the proportion of females produced by uninfected individuals. In fact, it is possible that the same cytogenetic mechanism operates in *Wolbachia* causing CI in diploids and haplodiploids; in diploids destruction of the paternal chromatin results in zygotic lethality whereas in haplodiploids it results in male production.

In theory, heritable symbionts will be selected to reduce the primary sex ratio of uninfecteds in diploid organisms. However, this has so far not been documented. How might it be accomplished? One scenario is as follows: symbionts present in testes produce DNA-binding proteins that associate with male- and female-determining genes in the developing sperm. Following fertilization by the sperm, the modifications (genomic imprinting) would then effect expression of the paternally derived sex determination genes during development of the zygote (i.e. sperm sex determination genes are imprinted). Specifically, the symbiont would be selected to increase expression of male-determining genes and decrease the expression of female-determining genes. To alter the sex ratio of uninfecteds *but not infecteds*, this modification would have to be rescued by the microorganisms in the egg (e.g. by increasing expression of female-determining genes). Symbiont-induced paternal sex-ratio modifications are shown in Fig. 1.2.

The scenario is not unreasonable, since microorganisms are already known to alter sex determination within infected hosts and to modify sperm function. A suggestive example occurs in *Glossina pallidipes*. Males infected with a vertically transmitted virus (when fertile) produce strong male-biased sex ratios when mated to uninfected females; however, uninfected females mated to uninfected males produce normal sex ratios (Jaenson 1986; Chapter 5). It remains to be seen whether symbionts pursuing the strategy of modifying sex determination through the male will be found. However, there is a simple and testable prediction for this strategy: uninfected females should produce more male-biased sex ratios when mated to infected males than when mated to uninfected males.

1.2.1.5 Increasing transmission rate

Heritable symbionts will almost always be selected to increase the vertical (maternal) transmission rate (a). The exception is when increasing transmission rate has negative pleiotropic effects on other adaptive phenotypes sufficient to offset the advantage of greater transmission. This point is made for cytoplasmic incompatibility microbes by Turelli (1994). For mutualistic symbionts, there will be strong selection on both the symbiont and the hosts to increase vertical transmission. Many of the transmission mechanisms for mutualistic symbionts described by Buchner (1965) reflect selection for stable transmission of mutualistic symbionts. Dunn *et al.* (1995) discuss some of the

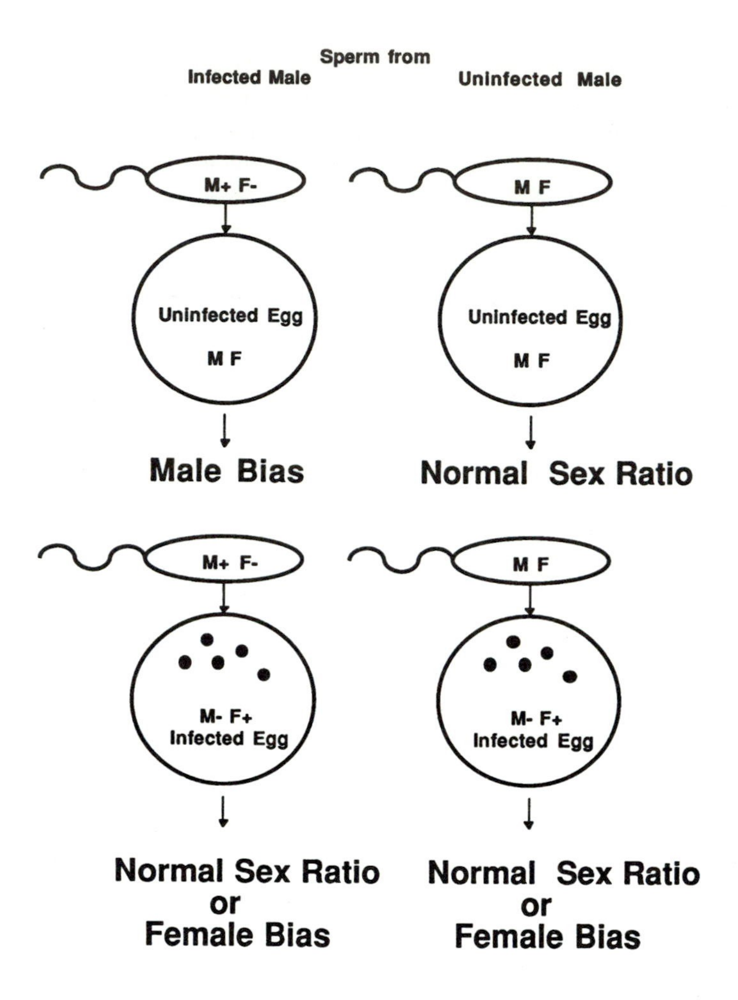

Figure 1.2 Symbiont-induced paternal sex-ratio modification. Symbionts present in males can be selected to decrease the proportion of daughters produced in matings with uninfected females. This could be achieved by imprinting sex determination genes during spermatogenesis, such that male-determining genes have greater expression (M+) and female-determining genes have reduced expression (F−). Symbionts present in eggs will be selected to compensate for these effects by increasing expression of female-determining genes and decreasing expression of male-determining genes.

problems relating to transmission for symbionts, particularly those that occur at low abundance within the host, such as feminizing protozoa in the amphipod *Gammarus duebeni*.

1.2.2 Horizontal and paternal transmission

Above, we have considered the adaptive strategies available to heritable symbionts with only vertical transmission. However, many microorganisms use a mixture of vertical and horizontal transmission. Microbes with a significant horizontal transmission component can readily evolve pathogenicity to the host (May and Anderson 1983). The human disease Rocky Mountain spotted fever is caused by an intracellular bacterium, *Rickettsia rickettsii*. The bacterium is transmitted to humans by ticks, and ticks can be reinfected from humans; however vertical transmission in ticks plays an important role in maintenance of the bacteria in tick populations (Burgdorfer 1975). Other examples of human diseases with both a horizontal and vertical transmission component include LaCrosse virus, dengue virus, hepatitis B and *Trypanosoma cruzi*, causative agent of Chagas disease (Busenberg and Cooke 1993).

The dynamics of disease-causing microorganisms with mixed horizontal and vertical transmission have been modelled extensively (May and Anderson 1983; Busenberg and Cooke 1993; Lipsitch *et al.* 1995); however, these models do not incorporate opportunities for sex-ratio distortion or cytoplasmic incompatibility. So far, no model is available that explores the dynamics of a symbiont with mixed transmission modes and also incorporates the full spectrum of possible reproductive strategies (i.e. sex-ratio distortion and cytoplasmic incompatibility).

Even a relatively low level of horizontal transmission can dramatically change the population dynamics of a symbiont, and coevolutionary dynamics with the host. For example, a horizontal component can sustain a symbiont within a population even when the survival of infected hosts is reduced (Lipsitch *et al.* 1995). Although *Wolbachia* show predominantly heritable transmission within species, their widespread distribution in arthropods is clearly due to horizontal (intertaxon) transmission (ONeill *et al.* 1992; Werren *et al.* 1995*a*).

Horizontal transmission can also allow for the mixing of symbiont lineages. Thus, if the symbiont is capable of sexual recombination, then horizontal transmission can increase the rate of adaptive evolution and can retard or prevent mutation accumulation via Muller's ratchet. Indeed, without some form of recombination between symbiont lineages, the rate of adaptation of symbionts to the host will be retarded and the rate of deleterious mutation accumulation enhanced.

1.3 Diversity of arthropod-associated symbionts

Arthropod symbionts can be grouped according to their location and degree of association with the host. These groupings include the reproductive endosymbionts which are localized to host gonad tissue (the main topic of this book), obligate nutritional endosymbionts (often associated with specialized host tissues adjacent to the gut), obligate extracellular nutritional symbionts found in the gut lumen, and facultative endosymbionts (often gut-associated

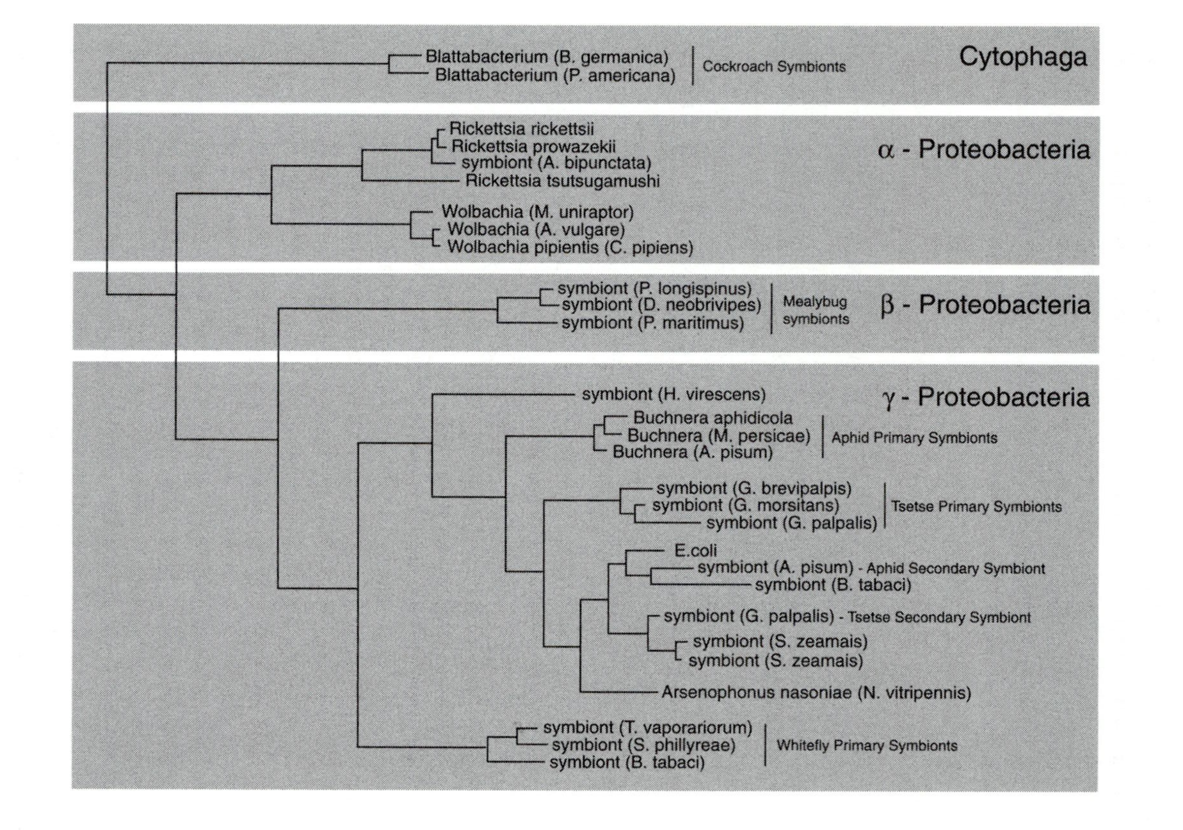

Figure 1.3 Phylogenetic tree derived from maximum parsimony analysis of aligned 16S rRNA sequences. The arthropod host is indicated after the bacterial name.

but not found in specialized tissues). The majority of symbionts that have been studied in detail are bacterial (and so they will receive greater attention in this book than other groups) but an extremely wide diversity of organisms is known to associate with arthropods, including inherited viruses, extracellular archaebacteria, intracellular yeasts, and extracellular fungi and protists. Taken as a group, it is clear that the formation of both intracellular and extracellular symbioses with arthropods has evolved many times independently (Fig. 1.3), involving a diverse array of microorganisms and arthropods.

1.3.1 Taxonomy and phylogeny of microbial symbionts

Microbiology has undergone a revolution in recent years as a direct result of the application of DNA and RNA sequencing to microbial systematics. This revolution has been most keenly felt by scientists working with fastidious microorganisms. Traditional microbial taxonomy was based on morphological and physiological characters which were inadequate to determine phylogenetic relationships between taxa with any confidence, particularly of those organisms that were unable to be cultured on cell-free media (Woese 1987, 1994). The result was a very superficial understanding of the taxonomy and phylogenetic relationships of fastidious microorganisms, which included most arthropod symbionts.

The application of DNA-based methodologies, in particular polymerase chain reaction (PCR) and sequencing of 16S rRNA genes, can provide an easily obtainable set of phylogenetic characters from any microorganism, including crudely prepared samples of whole-insect DNA which contain intracellular bacteria. The 16S rRNA gene has now become a standard molecule in microbial systematics. Sequences from this gene from a wide variety of eubacteria have been obtained and used for phylogenetic reconstruction. These studies have often provided results which have been at odds with prior taxonomy based on more traditional methods. These incongruencies reflect the inadequacies of the characters used in the past for classification. This is particularly true with regard to fastidious micro-organisms. For example, the currently accepted nomenclature for the genus *Wolbachia* lists three species within the genus: *W. pipientis*, *W. persica*, and *W. melophagi*. 16S analysis shows that *W. pipientis* and *W. persica* are quite unrelated bacteria—*W. pipientis* belonging to the α-Proteobacteria, together with members of the genus *Rickettsia* (ONeill *et al.* 1992), while *W. persica* belongs to the γ-Proteobacteria, most closely aligning with members of the genus *Francisella* (Weisburg *et al.* 1989). These discrepancies are proving to be extremely common, and most likely reflect convergent evolution of different bacteria to the intracellular life style.

While molecular data have been a boon for many workers in this field, a number of problems have not yet been adequately resolved. For example, a meaningful species concept that can be applied to microorganisms that are largely asexual is particularly troublesome and, as a result, the application of DNA sequence data in delimiting species boundaries is unclear at present. There is no obvious relationship between the degree of 16S sequence divergence and taxonomic status (Fox *et al.* 1992). For many bacteria, other physiological or ecological characters can readily be used to help delimit species boundaries, but for many intracellular symbionts, characters which may be meaningful in such an analysis are difficult to obtain, and so we are left with little more than DNA sequence data with which to classify many microbes. As a result, species determinations are extremely subjective and the criteria used to delimit species often vary from investigator to investigator.

One of the outcomes of molecular systematics as it has been applied to microbiology is that the new awareness of phylogenetic relationships is generating a new nomenclature for microbial taxonomy. This is a source of some confusion for non-microbiologists. In this book we will use the emerging classification and associated nomenclature that is arising from rRNA analysis (Table 1.1).

1.3.1.1 Reproductive symbionts

Microorganisms known to cause reproductive alterations within their arthropod hosts can be placed into two general categories. The *Wolbachia pipientis* assemblage is a widespread group of related α-Proteobacteria known to induce cytoplasmic incompatibility, parthenogenesis, and feminization in hosts (Breeuwer *et al.* 1992; ONeill *et al.* 1992; Rousset *et al.* 1992*a*; Stouthamer *et al.* 1993). For simplicity, we will refer to this group as the *Wolbachia*, excluding the unrelated *Wolbachia persica* and *Wolbachia melophagi*. Other reproduction-distorting symbionts include a collection of microorganisms of diverse origins that infect reproductive tissue of host arthropods. Most of these are known to cause some form of sex-ratio distortion, including host feminization or male-killing.

Wolbachia pipientis *and relatives*

The *Wolbachia* group has been known for many years, being first described from light microscopy analysis of mosquito ovaries in the 1920s (Hertig and Wolbach 1924). However, the connection between *Wolbachia pipientis* and reproductive phenotypes in its host, *Culex pipiens*, was not noticed until much later, in the 1970s (Yen and Barr 1971). In the past 10 years it has become apparent that this group is extremely widespread in arthropods, with PCR-based surveys indicating that around 15 per cent of all insect species may be infected with this agent (Werren *et al.* 1995b). Moreover, the use of 16S rRNA gene sequences has shown that a number of at first seemingly unrelated reproductive phenotypes are connected with *Wolbachia* infections in different arthropod species, including parthenogenesis (Chapter 4), cytoplasmic incompatibility (arising from incorrect sperm function during fertilization) (Chapter 2), modification of chromosomal sex determination in isopods (Chapter 3), and a possible influence on sperm competition (Wade and Chang 1995).

The *Wolbachia* assemblage is the sister group to various *Ehrlichia* species and is also closely related to the genus *Rickettsia* (ONeill *et al.* 1992). The extent of the host range of *Wolbachia* has yet to be fully established, but it is clear that it infects representatives of all the major insect groups as well as mites and isopods (Table 1.2). It has yet to be identified from ticks despite the clear propensity of this group to acquire bacterial symbionts. Interestingly, there is some evidence that *Wolbachia* may also be able to infect nematodes (Sironi *et al.* 1995). *Wolbachia* have been documented in over 98 invertebrate species by a combination of microscopy, genetic crossing, antibiotic curing, and molecular methods (PCR and sequencing). The current list of species known to harbour *Wolbachia*, and the methods used, is presented in Table 1.2.

Two major subdivisions of the *Wolbachia* which diverged from each other approximately 50 million years ago can be recognized, based upon synonymous substitution rates in the protein-coding *ftsZ* gene (Fig. 1.4; Werren *et al.* 1995*a*). It is clear from *Wolbachia* phylogeny as determined by both *ftsZ* and 16S rRNA sequences that there is no concordance with host phylogeny (Fig. 1.4), suggesting that these bacteria must utilize a combination of horizontal as well as vertical inheritance patterns (ONeill *et al.* 1992; Rousset *et al.* 1992*a*; Werren *et al.* 1995*a*,*b*). However, to date very little is known about the frequency and mode of horizontal transmission. Phylogenetic evidence indicates that one branch of *Wolbachia* (the A subdivision) has undergone extensive intertaxon transfer within the past 0–2.5 million years, although horizontal transfers occur within the B subdivision as well. Evidence also suggests that transfer of *Wolbachia* between parasitic insects and their hosts can occur (Werren *et al.* 1995*a*). Furthermore, there is little concordance between *Wolbachia* phylogeny based on 16S rRNA or *ftsZ* gene sequences and reproductive phenotypes generated in host insects. In other words, the *Wolbachia* strains that cause CI, for example, are not monophyletic with respect to the *Wolbachia* strains that cause parthenogenesis. Even within the CI *Wolbachia*, existing sequence data do not group together *Wolbachia* strains that generate the same crossing types (Rousset *et al.* 1992*b*; Werren *et al.* 1995*a*). These incongruencies could be explained in a number of ways, including the independent evolution on multiple occasions of the ability to generate different reproductive phenotypes; the possibility that host genome is the critical determinant of which reproductive phenotype is expressed; or that the *Wolbachia* genes that encode the molecules which cause the reproductive phenotypes are not located on *Wolbachia* chromosome but are extrachromosomal and so inherited independently of the *Wolbachia* chromosomal genes

Table 1.1 16S rRNA-based classification of prokaryotes (archaebacteria and eubacteria) showing the major divisions and groupings (after Maidak *et al.* 1994). Shaded groups contain intracellular inherited bacterial symbionts from arthropods

Domain/major divisions	
Eucaryota	
Archaea	
Euryarchaeota	
Methanococcales	
Methanobacteriales	
Methanomicrobacteria and relatives	
Methanomicrobiales	
Methanosarcinales	
Extreme halophiles	
Thermoplasmales and relatives	
Archaeoglobales	
Thermococcales	
Crenarchaeota	
Crenarchaeota group I	
Crenarchaeota group II	
Planktonic	
Xenarchaea	
Bacteria	
Thermophilic oxygen reducers	
Thermotogales	
Green non-sulphur bacteria and relatives	
Chloroflexus subdivision	
Deinococcus–Thermus subdivision	
Thermus group	
Deinococcus	
Flexibacter–Cytophaga–Bacteroides	
Subdivision I	
Bacteroides group	
***Cytophaga* group**	**Cockroaches, Termites**
Subdivision II	
Sphingobacterium group	
Saprospira group	
Flx. flexilis group	
Flx. litoralis group	
Cy. diffluens group	
Thermonema	
Rhodothermus	
Green sulphur bacteria	
Planctomyces and relatives	
Planctomyces subdivision	
Chlamydia subdivision	
Verrucomicrobium subdivision	

Taxon	Heritable symbionts
Cyanobacteria and chloroplasts	
Cyanobacteria	
Oscillatoria group	
Gloeothece gloeocapsa group	
Anabaena group	
Plectonema group	
Synechcococcus group	
Chloroplasts and cyanelles	
Fibrobacter	
Fibrobacter subdivision	
Fibrobacter group	
Acidobacterium subdivision	
Spirochetes and relatives	
Serpulina subdivision	
Spirochaeta–Treponema–Borrelia subdivision	
Spi. halophila group	
Treponema group	
Spi. aurantia group	
Borrelia group	
Leptospira subdivision	
Leptospira group	
Leptonema group	
Proteobacteria	
Alpha subdivision	
R. rubrum assemblage	
***Rickettsia* and others**	***Wolbachia* (insects, mites, isopods), ladybeetle male-killer**
Rhodobacter group	
Hyphomonas group	
Sphingomonas group	
Caulobacter group	
Rhizobium–Agrobacterium group	
Beta subdivision	
Neisseria group	
***Rub. gelatinosus* group**	**Mealybug symbionts**
Spr. volutans group	
Rhodocyclus group	
Nitrosomonas group	
Methylophilus group	
Gamma subdivision	
Ectothiorhodospira assemblage	
Chromatium group	
Sulphur-oxidizing symbionts	
Xanthomonas group	
Cardiobacterium group	
***Thiomicrospira* group**	***Wolbachia persica* (ticks)**
Legionella group	
***Methylomonas* group**	***Heliothis* symbiont**
***Oceanospirillum* group**	**Whitefly (secondary symbionts)**
Pseudomonas and relatives	
Colwellia assemblage	
Alteromonas group	
Vibrio group	

(Table 1.1 cont.)

Domain/major divisions	
Aeromonas group	
Enterics and relatives	**Aphid (P & S), *Camponotus* ants, tsetse fly (P & S), wasp son-killer, whitefly (primary symbionts)**
Haemophilus pasteurella	
Delta subdivision	
Desulfovibrio group	
Desulfuromonas group	
Bde. bacteriovorus group	
Desulfobulbus assemblage	
Myxobacteria	
Bde. stolpii group	
Epsilon subdivision	
Helicobacter and relatives	
Campylobacter and relatives	
Fusobacteria and relatives	
Gram-positive	
High G+C subdivision	
Atopobium group	
Streptomyces and relatives	
Frankia	
Actinomyces group	
Arthrobacter group	
Propionibactium group	
Catenuloplanes and others	
Saccharopolyspora group	
Mycobacterium complex	
Clostridia and relatives	
C. leptum group	
Thermoanaerobacter group	
C. butyricum group	
C. lituseburense group	
C. coccoides group	
C. purinolyticum group	
Syntrophomonas group	
Anaerobic halophiles	
Mycoplasmas and relatives	
M. hominis group	
M. pneumoniae group	
***Spiroplasma* group**	**Lady beetle male-killers, *Drosophila* male-killers**
Acholeplasma–Anaeroplasma group	
Bacillus–Lactobaciluus–Streptococcus subdivision	
Streptococci	
Lactobacilli	
Enterococcus assemblage	
Carnobacterium group	
Aerococcus group	
Listeria–Brochothrix group	
Planococcus group	

B. cereus group
Staphylococcus group
B. subtilis group
B. megaterium group
Sporolactobacillus group
B. brevis group
B. polymyxa group
Alicyclobacillus group

which have been sequenced so far. Further work is needed to discriminate between these different possibilities.

Reproductive distorters other than Wolbachia

Several quite unrelated symbionts have been reported to generate reproductive phenotypes in their hosts. These phenotypes are predominantly interactions with host sex determination or sex-specific lethality. The polyphyletic nature of host and symbiont combinations that can lead to sex-ratio distortions is a reflection of the diversity and labile nature of sex determination mechanisms in arthropods, often being quite sensitive to environmental conditions. Sex-ratio distortion has been reported from various insects infected with *Wolbachia* as well as from mites infected with *Rickettsia tsutsugamushi* (Roberts *et al.* 1977), and a number of older reports have incriminated both yeast-like intracellular symbionts and bacterial symbionts in the distortion of mite sex ratios (e.g. Buchner 1954, 1955), but as yet the mechanism underlying these observations is unclear. In addition, a protozoal symbiont of amphipods is known to alter sex determination toward females (Dunn *et al.* 1993*a*). Sex-specific lethality has been observed more commonly than direct interactions with sex determination mechanisms and, again, a variety of microorganisms have been implicated, including spiroplasmas in *Drosophila* species (Ebbert 1991), γ-Proteobacteria in parasitic wasps (Werren *et al.* 1986; Gherna *et al.* 1991), and *Rickettsia* in ladybirds (Werren *et al.* 1994) (see Table 1.1). These associations are discussed in detail in Chapters 3 and 5.

1.3.1.2 Mutualistic symbionts

Many symbiotic associations between arthropods and microbes occur in hosts that live on restricted diets. As a result it has often been assumed that the symbionts play a nutritive role for the host, helping to overcome the problems of a deficient diet. Many of these associations are obligate; if the symbiont is removed, the host will fail to complete its life cycle by being unable to successfully reach the adult stage or to be sterile if it does. Obligate nutritive symbionts may be extracellular in the gut lumen or intracellular within the gut epithelium, fat body, or in specialized cells known as bacteriocytes or mycetocytes.

Table 1.2 Known arthropod hosts of *Wolbachia*, phenotype associated with the infection where known and method by which the infection was detected

Species infected	Phenotype	Microscopy	Crossing/curing	PCR/sequencing
Phylum Nematoda				
Class Phasmidia				
Spirurida				
Filariidae				
Dirofilaria immitis				Sironi *et al.* (1995)
Phylum Arthropoda				
Class Crustacea				
Isopoda				
Armadillidiidae				
Armadillidium vulgare	Feminization	Rigaud *et al.* (1991*a*)	Martin *et al.* (1973)	Rousset *et al.* (1992*a*)
Armadillidium nasatum	Feminization	Juchault and Legrand (1979)	Juchault and Legrand (1979)	Rousset *et al.* (1992*a*)
Armadillidium album	Feminization	Juchault and Legrand (1979)		Bouchon (pers. comm.)
Oniscidae				
Chaetophiloscia elongata	Feminization	Juchault *et al.* (1994)		Juchault *et al.* (1994)
Ligiidae				
Ligia oceanica	Feminization	Juchault *et al.* (1974)		Bouchon (pers. comm.)
Porcellionidae				
Porcellio dilatatus	CI		Legrand *et al.* 1986)	Rousset *et al.* (1992*a*)
Porcellionides pruinosus	Feminization	Juchault *et al.* (1994)	Juchault *et al.* (1994)	Juchault *et al.* (1994)
Sphaeromatidae				
Sphaeroma rugicauda	Feminization	Martin *et al.* (1994)	Martin *et al.* (1994)	Martin *et al.* (1994)

Class Arachnida				
Acari				
Phytoseiidae				
Metaseiulus occidentalis				Johanowicz and Hoy (1996)
Phytoseiulus persimilis				Breeuwer and Jacobs (1996)
Galendromus occidentalis				Breeuwer and Jacobs (1996)
Neoseiulus barkeri				Breeuwer and Jacobs (1996)
Neoseiulus bibens				Breeuwer and Jacobs (1996)
Tetranychidae				
Tetranychus kanzawai				Breeuwer and Jacobs (1996)
Tetranychus neocaledonicus				Breeuwer and Jacobs (1996)
Tetranychus urticae				Johanowicz and Hoy (1996)
Tetranychus turkestani				Johanowicz and Hoy (1996)
Oligonychus biharensis				Breeuwer and Jacobs (1996)
Eutetranychus orientalis				Breeuwer and Jacobs (1996)
Class Insecta				
Coleoptera				
Chrysomedlidae				
Acromis sparsa				Werren *et al.* (1995*b*)
Chelymorpha alternans				Werren *et al.* (1995*b*)
Chersinellina heteropunctata				Werren *et al.* (1995*b*)
Diabrotica v. virgifera				O'Neill *et al.* (1992)
Cleridae				
Priocera sp.				Werren *et al.* (1995*b*)
Curculionidae				
Aramigus tesselatus				Werren *et al.* (1995*b*)
Bangasternus orientalis				Werren *et al.* (1995*a*)
Cossonus sp.				Werren *et al.* (1995*b*)
Hypera postica	CI	Hsiao and Hsiao (1985)	Blickenstaff (1965)	O'Neill *et al.* (1992)
Sitophilus oryzae				Werren *et al.* (1995*a*)

Table 1.2 (Cont.)

Species infected	Phenotype	Microscopy	Crossing/curing	PCR/sequencing
Dermestidae				
Attagenus unicolor				O'Neill *et al.* (1992)
Tenebrionidae				
Tribolium confusum	CI	O'Neill (1989)	Wade and Stevens (1985)	O'Neill *et al.* (1992)
Diptera				
Calliphoridae				
Protocalliphora sp.				Werren *et al.* (1995*a*)
Culicidae				
Culex pipiens complex	CI	Hertig (1936)	Ghelelovitch (1952) Yen and Barr (1973)	O'Neill *et al.* (1992)
Culex tigripes		Ndiaye and Mattei (1993)		
Aedes albopictus	CI	Wright and Barr (1980)	Kambhampati *et al.* (1993)	O'Neill *et al.* (1992)
Aedes brelandi			Taylor and Craig (1985)	
Aedes cooki	CI	Wright and Barr (1980)	Macdonald (1976)	
Aedes hebrideus	CI	Meek (1984)		
Aedes kesseli	CI	Wright and Wang (1980)	Trpis *et al.* (1981)	
Aedes malayensis	CI	Beckett *et al.* (1978)	Tesfa-Yohannes and Rozeboom (1974)	
Aedes polynesiensis	CI	Beckett *et al.* (1978)	Tesfa-Yohannes and Rozeboom (1974)	
Aedes pseudoscutellaris	CI	Meek (1984)	Meek and Macdonald (1984)	
Aedes riversi	CI	Wright and Barr (1980)		
Aedes s. scutellaris	CI	Meek (1984)	Smith-White and Woodhill (1954)	
Aedes tabu	CI	Beckett *et al.* (1978)		

Drosophilidae				
Drosophila ananassae	CI		Bourtzis *et al.* (1996)	Bourtzis *et al.* (1994)
Drosophila auraria	CI		Bourtzis *et al.* (1996)	Bourtzis *et al.* (1996)
Drosophila mauritiana	No phenotype		Giordano *et al.* (1995)	Rousset and Solignac (1995)
Drosophila melanogaster	CI—some strains no phenotype	King (1970)	Hoffman (1988)	Holden *et al.* (1993)
Drosophila quinaria				Werren and Jaenike (1995)
Drosophila recens	CI		Werren and Jaenike (1995)	Werren and Jaenike (1995)
Drosophila sechellia	CI		Giordano *et al.* (1995)	Rousset and Solignac (1995)
Drosophila simulans	CI—some strains no phenotype	Binnington and Hoffmann (1989)	Hoffmann *et al.* (1986)	Rousset *et al.* (1992*b*)
Drosophila tropicalis				Werren *et al.* (1995*b*)
Drosophila willistoni				Werren *et al.* (1995*b*)
Glossinidae				
Glossina austeni				O'Neill *et al.* (1993)
Glossina brevipalpis				O'Neill *et al.* (1993)
Glossina m. morsitans				O'Neill *et al.* (1993)
Neridae				
Nerius sp.				Werren *et al.* (1995*b*)
Psychodidae				
Phlebotomus papatasi				O'Neill (pers. comm.)
Stratiomyidae				
Cynomyia cyanea				Werren *et al.* (1995*b*)
Tephritidae				
Rhagoletis cerasi	CI		Boller *et al.* (1976)	
Rhagoletis pomonella				O'Neill *et al.* (1992)
Rhagoletis mendax				O'Neill *et al.* (1992)
Anastrepha suspensa				O'Neill *et al.* (1992)
Hemiptera				
Reduviidae				
Rhodnius pallescens				Werren *et al.* (1995*b*)
Rhodnius robustus				Hypsa (pers. comm.)

Table 1.2 (Cont.)

Species infected	Phenotype	Microscopy	Crossing/curing	PCR/sequencing
Delphacidae				
Laodelphax striatellus	CI	Noda (1984*b*)	Rousset *et al.* (1992*a*)	
Hymenoptera				
Agoanidae				
Tetrapus costaricensis				Werren *et al.* (1995*b*)
Aphelinidae				
Aphytis lingnanensis	Parthenogenesis		Zchori-Fein *et al.* (1995)	
Aphytis diaspidis	Parthenogenesis		Zchori-Fein *et al.* (1995)	
Aphytis yanonensis				Werren *et al.* (1995*a*)
Encarsia formosa	Parthenogenesis		Zchori-Fein *et al.* (1992)	Werren *et al.* (1995*a*)
Apidae				
Trigona sp.				Werren *et al.* (1995*b*)
Braconidae				
Asobara tabida			Werren *et al.* (1995*a*)	
Cynipidae				
Diplolepis rosae				van Meer *et al.* (1995)
Encyrtidae				
Apoanagyrus diversicornis			Pijls *et al.* (1996)	Werren (pers. comm.)
Eulophidae				
Mellitobia sp.				Werren *et al.* (1995*a*)
Eucoilidae				
Leptopilina australis	Parthenogenesis		van Alphen (pers. comm.)	Werren *et al.* (1995*a*)
Leptopilina clavipes	Parthenogenesis		van Alphen (pers. comm.)	Werren *et al.* (1995*a*)
Formicidae				
Ectatomma tuberculatum				Werren *et al.* (1995*b*)

Proctotrupoidae				
Trichopria drosophilae	CI		van Alphen (pers. comm.)	Werren *et al.* (1995*a*)
Pteromalidae				
Muscidifurax uniraptor	Parthenogenesis		Stouthamer *et al.* (1994)	Stouthamer *et al.* (1993)
Nasonia vitripennis	CI		Saul (1961)	Breeuwer *et al.* (1992)
Nasonia giraulti	CI		Breeuwer and Werren (1990)	Breeuwer *et al.* (1992)
Nasonia longicornis	CI			Breeuwer *et al.* (1992)
Spalangia fuscipes				Werren *et al.* (1995*a*)
Sphecidae				
Tropoxylon sp.				Werren *et al.* (1995*a*)
Trichogrammatidae				
Trichogramma brevicapillum	Parthenogenesis	Stouthamer and Werren (1993)	Stouthamer and Werren (1993)	Werren *et al.* (1995*a*)
Trichogramma chilonis	Parthenogenesis		Stouthamer *et al.* (1990*a*)	
Trichogramma cordubensis	Parthenogenesis	Stouthamer and Werren (1993)	Stouthamer *et al.* (1990*a*)	Rousset *et al.* (1992*a*)
Trichogramma deion	Parthenogenesis	Stouthamer and Werren (1993)	Stouthamer *et al.* (1990*a*)	Stouthamer *et al.* (1993)
Trichogramma embryophagum	Parthenogenesis	Stouthamer and Werren (1993)	Stouthamer *et al.* (1990*a*)	
Trichogramma evanescens	Parthenogenesis	Stouthamer and Werren (1993)	Stouthamer *et al.* (1990*a*)	
Trichogramma nr deion	Parthenogenesis		Stouthamer and Kazmer (1994)	
Trichogramma oleae	Parthenogenesis	Louis *et al.* (1993)	Stouthamer *et al.* (1990*a*)	Rousset *et al.* (1992*a*)
Trichogramma platneri	Parthenogenesis	Stouthamer and Werren (1993)	Stouthamer *et al.* (1990*a*)	
Trichogramma pretiosum	Parthenogenesis	Stouthamer and Werren (1993)	Stouthamer *et al.* (1990*a*)	Stouthamer *et al.* (1993)

Table 1.2 (Cont.)

Species infected	Phenotype	Microscopy	Crossing/curing	PCR/sequencing
Lepidoptera				
Nymphalidae				
Cissia usitata				Werren *et al.* (1995*b*)
Cissia libye				Werren *et al.* (1995*b*)
Pyralidae				
Parapoynx sp.				Werren *et al.* (1995*b*)
Cadra (=*Ephestia*) *cautella*	CI	Kellen *et al.* (1981)	Brower (1976)	O'Neill *et al.* (1992)
Ephestia kuehniella				Rousset *et al.* (1992*a*)
Corcyra cephalonica				O'Neill *et al.* (1992)
Orthoptera				
Gryllidae				
Gryllis pennsylvanicus				Werren *et al.* (1995*a*)
Tettigoniidae				
Lophaspis scabricula				Werren *et al.* (1995*b*)

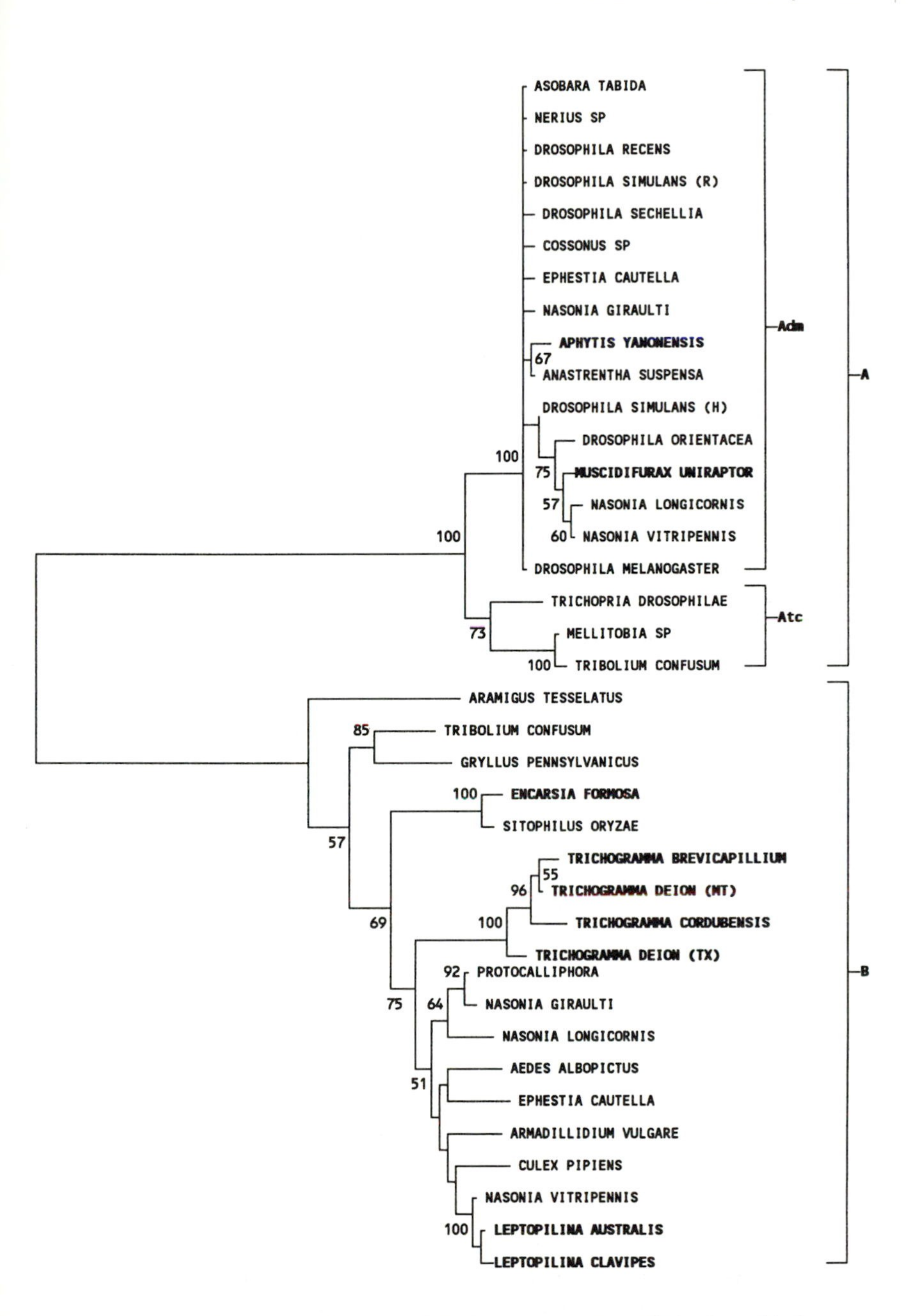

Figure 1.4 Phylogenetic tree of *Wolbachia* based upon sequences of the bacterial *ftsZ* gene. Parthenogenesis-associated bacteria are shown in bold. The tree was generated by neighbour-joining using the *p*-distance including insertions and deletions. Numbers next to the nodes indicate the number of replicates confirming the node out of 100 (only numbers greater than 50 are shown). The tree clearly shows horizontal transmission of *Wolbachia* between different orders of insects. Closely related *Wolbachia* are found in a parasitic wasp (*Nasonia giraulti*) and its fly host (*Protocalliphora*), suggesting horizontal transfer between parasite and host insects. (Redrawn from Werren *et al.* 1995*b*.)

Obligate relationships

Extracellular obligate symbiont associations have been studied in detail in a number of insect groups. For example, rich assemblages of various microorganisms are found in the hind-guts of termites and some cockroaches which have a predominantly cellulose diet. Indeed, termite guts have been described to contain representatives from the three domains of life, including bacteria, archaebacteria and eukaryotes (Ohkuma *et al.* 1995; Ohkuma and Kudo 1996). Similarly, insects that feed solely on vertebrate blood or plant sap are often associated with obligate symbionts. A well-studied example is the extracellular, Gram-positive *Rhodococcus* bacteria that infect triatomine bugs (Baines 1956).

Many obligate associations also involve intracellular symbionts, predominantly bacterial or yeast-like organisms. Again, similar life histories of restricted diets are good indicators of insects that contain these symbionts. For example, aphids, mealybugs, and whiteflies are known to contain intracellular symbionts associated with bacteriocytes whose presence is required by the host. Often these insects contain a number of different symbiont associations, but the obligate bacteriocyte-associated symbionts are known as 'primary' symbionts. Phylogenetic analysis of the 16S rRNA genes from these primary symbionts shows that they commonly have a concordant phylogeny with their host insects (Moran and Baumann 1994). This indicates that these obligate nutritive associations have, for a long time, been solely dependent on strict vertical inheritance. This is in contrast to the facultative reproductive symbionts mentioned above which show no such concordance. In addition to the hemipteran examples, similar patterns of evolution have also been found for primary symbionts in cockroaches (Bandi *et al.* 1994) and in tsetse flies (Aksoy *et al.* 1995). Indeed, tsetse flies have a unique set of symbiont associations. Three categories of symbionts have been described from these blood-feeding flies to date, including *Wolbachia*, primary bacteriocyte-associated bacteria, and secondary facultative gut-associated intracellular bacteria. Of these three associations, only the primary symbionts are needed for the fly to survive and reproduce, and again these bacteria (*Wigglesworthia glossinidia*) have a concordant evolution with their hosts, indicating strict vertical inheritance (Aksoy *et al.* 1995).

This same pattern is also seen with intracellular bacterial symbionts of cockroaches. These symbionts are localized to the fat body of the roaches and, although they are quite unrelated to the γ-Proteobacteria which commonly form these mutualistic associations with insects, they still display a similar concordant phylogeny with their hosts. Indeed, even the primitive termite, *Mastotermes*, contains bacterial endosymbionts of the same group, which show a deep branching relationship to the cockroach endosymbionts. This parallels the commonly held insect phylogeny that relates termites and cockroaches (Bandi *et al.* 1995).

Similarly, attine ants (including leaf-cutter ants) have associated symbiotic fungi which they cultivate on plant material collected into subterranean gardens, upon which they are dependent for nutrition. Phylogenetic studies indicate that primitive attines have repeatedly acquired their symbionts from free-living fungi, whereas the fungi of derived attines have concordant phylogenies with the host, indicating heritable transmission (Chapela *et al.* 1994). This is consistent with the observation in some higher attine species, that the female collects a sample of fungus prior to dispersal and founding of a new colony (Chapela *et al.* 1994). It is likely that the specialized ant-fungus symbiosis has been a motor of evolutionary change in both the ants and associated fungi.

Many of these associations must be extremely ancient. To date most of the studied primary symbionts have been bacterial, and predominantly members of the γ-Proteobacteria (Table 1.1). However many known associations have yet to be studied in detail, including the yeast-like primary symbionts which are commonly found in hemipterans (Buchner 1965; Noda *et al.* 1995) and some coleopterans (Noda and Kodama 1996).

Facultative relationships

In contrast to the strictly inherited obligate nutritive symbionts of many insects, many associations are facultative. A spectrum of agents falls into this grouping which shows the full range of abilities to be horizontally and vertically transmitted. Many of the studied examples are human pathogens, including the various *Rickettsia* species which are readily acquired orally by arthropods as well as being transmitted vertically (Azad *et al.* 1992). Here we will only consider the intracellular examples, although the majority of members of this class of association are probably extracellular. The best-studied examples are the intracellular secondary symbionts of aphids and tsetse flies (Beard *et al.* 1993). These bacteria are intracellular but not contained within bacteriocytes. In many cases the phenotypic consequence of these infections to the host is unknown and, owing to the difficulty in working with these systems, will likely remain difficult to determine. It is clear, however, that these agents are capable of horizontal transmission between hosts, as well as vertical transmission. The secondary symbionts of tsetse flies show no phylogenetic concordance with their hosts, unlike their primary counterparts (Aksoy *et al.* 1995, 1997). However, despite this lack of linkage between the symbionts and hosts, it is still probable that transmission in these cases is commonly vertical.

1.4 Potential conflicts between symbionts and hosts

Symbionts, whether they be primarily mutualists, sex-ratio distorters or cytoplasmic incompatibility microbes, do not have completely convergent interests with their hosts. There are at least three main arenas in which host and symbiont can have conflict:

(1) sex determination and sex ratio of the host;

(2) regulation of symbiont population density; and

(3) symbiont transmission and host germline determination.

1.4.1 Sex determination

The basic reasons for conflict over sex determination have already been mentioned; symbionts are typically inherited asymmetrically (through females but not through males). Hence males are a genetic dead-end for symbionts. In contrast, the nuclear genes (excepting heterochromatic sex chromosomes) are transmitted through both sexes. This creates a genetic (or intragenomic) conflict between nuclear genes and cytoplasmically inherited elements (such as mitochondria and symbionts) over sex determination and sex ratio of the host (Eberhard 1980; Cosmides and Tooby 1981). Symbionts are selected to convert nuclear genotypic males into females or to bias the sex ratio of mothers toward female production. However, this phenotype creates selection upon the nuclear genome to counteract these effects. Hurst *et al.* (Chapter 5), Rigaud (Chapter 3) and Stouthamer (Chapter 4) discuss interesting examples of microbial sex-ratio distorters, and Rigaud (Chapter 3) extensively considers the consequences of genetic conflict over sex determination and how it may have influenced the evolution of isopod sex determination.

An interesting area that needs further investigation is to what extent mutualistic symbionts have evolved mechanisms to alter host sex determination and to what extent hosts have evolved counterstrategies. As mentioned above, many heritable mutualistic symbioses are quite old. For example, symbiotic relationships between aphids and *Buchnera* are estimated to go back over 200 million years (Moran and Baumann 1994). Symbiotic relationships (presumably ancient) are common in coccids, involving in some cases bacteria and in others yeast. Other groups with common mutualistic symbioses include weevils, lice, cockroaches, and isopterans.

We predict that the genetic, biochemical and physiological architecture of sex determination will be influenced by conflict between these symbionts and their hosts. In general, symbionts will be selected to produce products that push sex determination towards females, whereas the host is expected to compensate by producing products that push sex determination towards males. Under most circumstances, we expect that the host will maintain ultimate control over sex determination, primarily by regulating the symbionts and compensating for their action. However, the architecture of sex determination will still be influenced by this conflict. Selection might also favour mutualistic symbionts that also cause male-killing, when there is competition among siblings for resources or inbreeding depression via sib mating. Interestingly, although mutualist variants that cause male-killing can be selectively favoured, their existence in high frequencies could lead to population (or species) extinction. Because male-killing microbes reduce host fitness, selection will act upon the host to suppress their action and to reduce opportunities for male-killing.

The role of mutualists in host sex determination has not been extensively investigated, and this promises to be an interesting area of pursuit. One might argue that obligatory mutualistic symbionts are already at fixation within their hosts, and therefore do not need to cause sex-ratio distortion. However, variants of mutualistic symbionts that also alter sex determination will be at a selective advantage relative to the more benign forms, for the reasons discussed above. Therefore, mutualistic symbionts that reside in the reproductive tissues and early embryos might produce products within eggs that effect sex determination. Targets of these products are expected to be genes or gene products involved in primary sex determination. Hosts would therefore be expected to sequester symbionts and to minimize their biochemical activities, particularly during these times. Sequestering of mutualistic symbionts during oogenesis and early development is a common theme observed in many hosts with mutualistic endosymbionts (Buchner 1965).

Clearly, the extent to which mutualistic symbionts can pursue these reproductive strategies depends upon the underlying genetic mechanism of sex determination (as pointed out by Rigaud, Chapter 3), and the ability of the microorganism to generate gene products which influence host sex determination. In species with male heterogamety (XY), where genes necessary for male determination are located on the Y (e.g. mammals), then simple feminization of XY males could be attainable by suppressing the male-determining genes. Such symbionts could spread to appreciable frequencies even if YY grand-progeny are inviable, particularly if they have high transmission rates (Taylor 1990). In species with female heterogamety (ZW females), symbionts can be strongly selected to both feminize ZZ males (as seen in *Armadillidium vulgare*) and to enhance meiotic drive of the W chromosome in female gametogenesis, even if this chromosome is heteromorphic. Similarly, XO male : XX female systems provide opportunities for feminizing and sex-ratio distorting mutualistic symbionts, by restoration of XX within eggs.

Several systems suggest opportunities for mutualistic symbionts to alter host sex determination. Most aphids are cyclical parthenogens. As described in White (1973), the typical developmental pattern is as follows. Females reproduce parthenogenetically for several generations, followed by parthenogenetic production of sexaparae, individuals capable of producing sexual males and females. The sexaparae produce two types of eggs, males and females. In the majority of species, males are XO whereas females are XX. Both type of eggs develop parthenogenetically and the autosomes undergo non-reductional (mitotic) divisions; however, in male eggs, the X chromosomes form a bivalent and one X segregates to the polar body. Thus, sex must be determined by factors placed into the egg during oogenesis that effect X chromsome behaviour. Buchner (1965) reports that symbionts are transmitted to both male and female eggs in most species. Aphid symbionts are expected to produce products that interfere with X chromosome loss and the host to counteract these effects. Similarly, aphid symbionts may produce products that sustain parthenogenetic reproduction, with host suppression of these effects during the sexual phase.

In some cases the possible footprint of male-killing symbionts is apparent. In some pemphigine aphids, symbionts are not present in male embryos but are in female embryos. In these species, the males are reduced in size and do not feed. It can be argued that symbionts are not needed in these species due to lack of male feeding (Buchner 1965); however, there is an alternative interpretation. In systems where male-killing mutualistic symbionts are selected for, suppression of the male-killing may be achieved by exclusion of the symbiont from male eggs. The cost of this exclusion will be frail males who are small and do not feed. Thus, these situations are most likely to arise where there is parental brooding (to provide sustenance to the males), conditions that also favour male-killing (Werren 1987; L. D. Hurst 1993). In several coccid groups (*Pseudococcus*, *Puto*, *Macrocerococcus*) the mycetomes in embryonic males are very small or minute (Buchner 1965). Males in these groups also do not feed and females brood their young (Nur, personal communication). Buchner (1965) also reports that in coccid species with yeast symbionts, the symbionts are excluded from male eggs but not female eggs. One mechanism by which hosts could avoid the negative effects of male-killing nutritive symbionts is to conceal the sex of the progeny. Interestingly in this regard, coccids which undergo paternal genome loss in males actually retain the paternal genome within the symbiont-bearing mycetomes (Brown 1965). The patterns described above are merely suggestive. Clearly, much remains to be done to determine whether mutualistic symbionts are actually involved in host sex determination and sex-ratio distortion.

1.4.2 Regulation of symbiont numbers

The regulation of symbiont numbers in hosts is poorly understood. From the evolutionary perspective, the interests of host and mutualistic symbionts are generally concordant. Excessive replication of symbionts is likely to reduce the fitness of hosts, and therefore that of the symbionts, since the host is the vehicle for their propagation to future generations. Thus, in general we will expect the evolution of prudent symbionts that have reduced replication rates so as not to significantly harm the host. However, evolutionary interests of the symbiont and host are not completely concordant.

In many respects the population genetics of heritable symbionts is analogous to that of mitochondria, a topic extensively investigated by Birky and colleagues (Birky 1978; Backer and Birky 1985; Birky *et al.* 1989). As with mitochondria, symbionts are inherited uniparentally through the maternal line. As a result, there is little or no mixing between symbiont lineages. A second feature in common with mitochondria is the hierarchical structure of symbiont populations. There is the individual symbiont, population of symbionts within individual host cells (nutritive symbionts are often localized in specialized cells, mycetocytes or bacteriocytes, until the time of host reproduction and transmission), population of infected host cells within an individual host organism, and populations of infected hosts. The population dynamics of symbionts will be dependent upon stochastic processes of transmission and selection at the different levels.

Selection on symbiont-induced phenotypes will occur at each of the organizational levels, and can sometimes be antagonistic at different levels. For example, both within-host selection and between-host selection act upon symbionts. Phenotypes that favour individual symbionts in competition within a host can be detrimental to the fitness of the host and associated symbiont lineages (Maynard Smith and Szathmary 1995). Consider a mutation that arises within an individual symbiont that confers a replication advantage (e.g. higher replication rate) relative to the other symbionts within a host cell. The frequency of the mutant will increase during successive replication cycles within the host cell or within the host cell lineage. Eventually the mutant can go to fixation within the host lineage, whether or not the symbiont variant is harmful to the host. Variants that have a replication advantage but are harmful to host fitness will be selected for by within-host selection but selected against by between-host selection.

Clearly, symbiont mutants can also have a replication advantage without being detrimental to the host, and such symbionts should become common in host populations. The rate at which this occurs will depend upon how readily mutations for higher replication rates occur and to what extent there is mixing of symbionts within and between hosts. Symbiont mutations that increase relative replication rates but are not harmful to the host can occur, for instance, when symbionts compete for a limiting resource provided by the host. Competition among symbionts within a host will be most intense at the time of transmission to progeny. Competition will occur for transmission to the eggs and for representation within the egg, thus favouring higher replication rates at this time. Interestingly in this regard, Buchner (1965) observed that, in general, a conspicuous spontaneous increase of the symbionts goes hand in hand with egg infection. This is particularly dramatic in *Camponotus* ants, where a stormy increase of symbionts temporarily results in a large proportion of the egg biomass being made up by symbionts. However, in general, symbiont mutations that are detrimental to the host cannot persist because they are at a competitive disadvantage relative to more benign forms occurring within other hosts (i.e. they are eliminated at the level of interhost selection).

1.4.2.1 Symbiont cancers

Both hosts and symbionts face an evolutionary problem. Symbiont mutations that are extremely harmful to the host but increase replication rate will increase within the host. These can be termed symbiont cancers. They will reduce the fitness of both the host and associated non-cancerous symbionts. Adaptations on the part of both the host and symbiont are expected to evolve to deal with this problem. Symbiont genome reorganizations that reduce the probability of rapidly replicating symbiont cancers should be selectively favoured (e.g. Rouhbakhsh and Baumann 1995; Bensaadimerchermek *et al.* 1996). Transfer of vital bacterial genes to the host genome could provide an additional mechanism for regulating symbiont replication. Although mitochondrial protein genes are known to have transferred to the host nuclear genome, it is not known how widespread this process is in non-organelle symbionts.

Hosts may directly regulate symbiont replication (and reduce the probability of symbiont cancers) by withholding limiting nutrients. Buchner (1965) observed that the nutrient conditions under which they (symbionts) live in the hosts is obviously sparse. It may seem incongruous that hosts would starve their symbionts; however, this arises from the divergence of interests of the two parties and the possibility of within-host selection favouring mutants for uncontrolled symbiont growth. Another mechanism for reducing within-host selection on symbionts is for the host to restrict the number of symbiont cells that are transmitted to progeny. By forcing the symbiont population through a bottleneck, all symbionts within an individual will be derived from a few related cells, thus reducing the possibility for competition between symbiont variants within a host (Maynard Smith 1991; Frank 1994). There is some evidence for this, again from Buchner (1965), who states that for some reason it is apparently to the advantage of host animals to localize the creation of transmission forms narrowly. However, a consequence of this is that deleterious mutations can accumulate in asexual symbiont lineages due to Muller's ratchet (Moran 1996).

In summary, different levels of selection acting upon symbionts could create selection for forms that are advantageous in the short term (within hosts or host lineages) but disadvantageous in competition between symbiont lineages. Therefore, we can expect that for long-term associations between mutualistic symbionts and hosts, selection will favour genetic reorganizations of symbionts that reduce the possibility of symbiont cancers, and hosts that restrain the opportunities for unfettered symbiont growth.

1.4.3 Germline transmission and germline determination

Selection will act strongly upon symbionts to increase their transmission through the germline. Clearly, in the case of mutualistic symbionts the host is also selected to transmit symbionts. However, the evolutionary interests of host and individual symbiont cells are not identical. Thus, mutations in symbionts that increase their germline transmission will be selectively favoured unless they significantly reduce the fitness of the host.

1.4.3.1 Germline transmission

Heritable mutualistic symbionts are transmitted to progeny via a number of different mechanisms which can be divided into three general categories: (a) oral uptake by young, often associated with contamination of the egg surface; (b) intra-egg (cytoplasmic) infection; or (c) embryonic infection. A review of these mechanisms is provided by Koch (1967). Triatomine bugs feed on faecal drops deposited by the mother, thus ensuring uptake of nutritive symbionts. Tsetse appear to transmit their secondary symbionts orally through milk gland secretions (ONeill *et al.* 1993). Reproductive symbionts can also show a similar transmission mode. The male-killing bacterium of *Nasonia* wasps (which sting and lay eggs within fly pupae) is transmitted to the fly haemolymph during stinging. The bacteria replicates within the fly pupae and then infects the feeding wasp larvae orally (Huger *et al.* 1985).

In many species with mutualistic extracellular gut symbionts, transmission is ensured by the mother smearing symbiont-contaminated material upon the egg surface. Gut symbionts are expected to increase replication and to enhance transmission at this time (e.g. perhaps by symbiont-induced diarrhoeas). For example, in the bug *Pentatoma rufipes*, there is a dramatic increase in symbiont replication during sexual maturation, resulting in an abundance of infective material (Koch 1967). Often, specialized structures are present for delivery of symbionts to the egg surface. Feeding by the hatching young on the egg case ensures transmission. Clearly, we would expect competition among the population of symbionts in the gut for transmission to these specialized structures. In addition, pathogens and parasites can also be expected to compete for transmission via these mechanisms.

Intracellular nutritive symbionts that are localized within specialized structures (mycetomes) around the gut must be transmitted to the developing eggs. Such symbionts either actively find their way to the egg or are transferred there by specialized cells or other structures (Koch 1967). Transmission is in the interests of both the host and symbionts. However, intense competition among symbionts for successful transmission might be expected. This competition should be partly ameliorated by the high degree of genetic relatedness among symbiont cells. However, as previously discussed, mutations that impart an advantage during transmission will be selectively favoured by within-host selection. Eventually, we expect the refinement of symbiont phenotypes that restrict competitive behaviour to these particular times, thus minimizing negative effects on host (and therefore symbiont lineage) fitness.

As previously described, bacterial symbionts in *Camponotus* and *Formica* ants are transmitted to the eggs and undergo high rates of replication that are far in excess of what would be necessary to merely ensure transmission (Buchner 1965; Koch 1967). The pattern is best explained in terms of intrahost competition among symbionts. In cicadas, at the time of ovarian development, the symbionts swarm out of the mycetomes and arrive at the place of infection by way of the lymph where they enter specialized wedge cells near the follicles prior to entering the egg (Koch 1967, p. 45). Again, although anecdotal, these accounts imply a much greater movement of symbionts than would be required merely to ensure transmission. In *Macrocerococcus superbus*, intact mycetocytes actually enter the developing egg and fuse with yolk cells in a kind of somatic fertilization (Koch 1967). This mechanism would reduce the potential for symbiont competition, depending upon how many different mycetocytes contribute to each egg.

For reproductive parasites, the evolutionary interests of host and symbiont are less likely to be concordant. The host is not necessarily selected to ensure symbiont transmission, and in some cases may be favoured to suppress it. For example, hosts infected with male-killing or other sex-ratio distorting microbes will generally be selected to reduce or eliminate transmission of the reproductive parasites (Uyenoyama and Feldman 1978; Werren 1987). For cytoplasmic incompatibility (CI) bacteria, the situation is more complex (Turelli 1994). During initial invasion of a population by CI bacteria, selection can favour host genotypes that reduce CI bacterial transmission. However, if the infection is near fixation within the host population, the host can actually be selected to enhance transmission of the bacteria to progeny (at least to daughters) because the daughters will then be reproductively compatible with infected males in the population. Reproductive symbionts themselves are under strong selection to enhance their transmission rates, so long as the cost to (female) host fertility is not too great. In this regard, the CI *Wolbachia* of *Nasonia* wasps localize to the pole of the egg where germ cells develop, thus presumably enhancing their transmission through the germline (Breeuwer and Werren 1990). This contrasts with CI *Wolbachia* in *Drosophila simulans* and *D. melanogaster*, where bacteria are distributed throughout the developing syncytial blastoderm (ONeill and Karr 1990), although they are found predominantly within the gonadal tissue of adults. This may reflect a more ancient association of *Wolbachia* in *Nasonia*, resulting in specific adaptations for germline transmission within that host.

1.4.3.2 Germline determination

Could symbionts be selectively favoured to influence germline determination in the developing embryo? Under some circumstances, the answer may be yes. First, it should be clear that symbiont mutants that induce gross abnormalities in germ-cell formation will generally not be selected for. Although such mutants may gain a short-term transmission advantage (i.e. greater representation among the gametes) in the first host generation, they are likely to cause sufficient detrimental effects upon host fitness to be selectively eliminated by interhost selection. However, more subtle influences on germline development may evolve.

Germ cells are determined early in development in most animals. Imagine the following situation. Suppose that germ cells are determined by a gradient of germline determinant products within the egg cytoplasm. Cells that form within the region containing germline determinants will develop into primordial germ cells. This is the pattern observed in a number of species. For example, in *Drosophila*, germline determinants called polar granules localize to the germpole of the egg and are composed of the products (both protein and mRNA) of several different genes (Lehmann and Ephrussi 1994). Germ-specific granules are also found in such diverse organisms as nematodes and frogs (Lehmann and Ephrussi 1994). Intracellular symbionts that are free within the host cytoplasm will be selected to localize within the general region containing germline determinants (as described above). However, there is likely to be a gradient in such determinants within the cytoplasm. Those symbionts in the periphery of the germ determinant region will be less likely to be incorporated within germ cells. A symbiont mutant that increases its probability of incorporation, either by interacting with or producing germ-cell determinants, will increase in frequency so long as such interactions do not dramatically disrupt host fitness.

1.5 Other evolutionary consequences of heritable symbiosis

1.5.1 Evolution of novel phenotypes

It is now widely recognized that the union of symbiont and host provides opportunities for the evolution of novel phenotypes, by combining genomes with different biochemical capabilities (Margulis and Fester 1991). Examples of this phenomenon abound, including the symbiotic origin of mitochondria and chloroplasts (Margulis 1981), sulphide-oxidizing bacteria found in various invertebrates in hydrothermal vents and other sulphide-rich environments (Vetter 1991), and a variety of nutritional endosymbionts in arthropods (Buchner 1965).

Not only does symbiosis allow organisms to occupy novel habitats due to an expanded biochemical repertoire, but the very interplay of coevolutionary dynamics between host and symbiont almost certainly accelerates evolution of both parties. It is quite possible that host taxa that acquire heritable symbionts have higher rates of morphological and physiological evolution than do equivalent taxa without symbionts. Similarly, such taxa may have higher rates of speciation. Systematic comparative studies to determine whether this is the case have not yet been undertaken. Two suggestive examples occur in ants. In general, nutritive symbionts are uncommon in ants. However, carpenter ants (genus *Camponotus*), which are among the most speciose of ant genera, harbour endosymbiotic bacteria. attines (fungus-ants) maintain mutualistic symbioses with fungi. The more derived attines have evolved heritable transmission patterns with their fungi and appear to show accelerated rates of speciation (Chapela *et al.* 1994; D. Mueller, personal communication).

Cytoplasmic incompatibility *Wolbachia* may also increase rates of speciation within hosts, by promoting reproductive incompatibility between incipient species. Several suggestive examples occur. Partial to complete bidirectional incompatibility has been found between geographic isolates of *Drosophila simulans* (ONeill and Karr 1990), geographic strains of *Culex pipiens* (Laven 1959) and sibling species of *Nasonia* wasps (Breeuwer and Werren 1990). In *Nasonia* wasps, *Wolbachia*-induced bidirectional incompatibility between species prevents the formation of hybrids in interspecies matings, at least under laboratory conditions (Breeuwer and Werren 1990). Interestingly, in this species segregation of different *Wolbachia* types among sublines of a single strain can rapidly lead to nearly complete reproductive isolation between sublines (Perrot-Minnot *et al.* 1996). It has yet to be demonstrated that *Wolbachia* promote rapid speciation, nor has reinforcement been demonstrated in conjunction with CI-induced reproductive isolation; however, the abundance of *Wolbachia* within arthropods (Werren *et al.* 1995*b*) suggests their potential importance.

1.5.2 The problem of symbiont function in males

As has been described, maternally inherited symbionts can be selected to kill males under certain special conditions, i.e. when the fitness of infected females from the same lineage is increased. However, there is a more subtle problem relating to symbionts and males that has not received widespread attention. Because cytoplasmically inherited symbionts are not transmitted through males, *there is no direct selection upon symbionts to maintain function within males*. As a consequence, any mutations arising within symbionts that reduce (or eliminate) performance in males will not be at a selective disadvantage. Such mutations can spread to fixation within the symbiont population by drift, selection (for performance in females) or hitch-hiking. This can occur even if the symbiont causes major deleterious effects in males, since there is no counteracting selective pressure.

This effect could have important consequences. For example, the absence of symbionts within males of some species, and concomitant reduction in male size and vigour, may be a consequence of poor symbiont performance in males. Similarly, the retention of paternal chromosomes in coccids, previously mentioned as possible evidence for male-killing symbionts, could also be due to the lack of adaptation of symbionts to haploid (male) tissues in these species. For many symbiont functions, there may be little difference between performance within male cells versus female cells. However, in cases where male problems do arise, the host genome could evolve to adjust to changes in symbionts that interfere with male performance.

The problem of males has an interesting manifestation in cytoplasmic incompatibility bacteria. These bacteria are found in both testes and eggs. Bacteria in testes apparently modify the sperm (or sperm chromatin) in such a way that the sperm dysfunctions if it fertilizes an egg in which the bacteria are not present to rescue the modification (Chapter 2). The bacteria can increase in host populations because they reduce the fitness of hosts which do not carry the bacteria. However, there is no direct selection to maintain function within males. The benefit of reducing frequency of uninfecteds in the population will be gained by all bacterial lineages capable of rescue within the eggs, whether or not they induce modification within male sperm. Thus male-defective CI bacteria can be selectively favoured in populations with functional CI bacteria; however, they cannot increase in frequency in the absence of normal CI strains (Turelli 1994; Hurst and McVean 1996). Thus, the male-defective mutants can be considered as parasites upon the functional forms.

1.5.3 Evolution to organelles

The evidence is now overwhelming that both mitochondria and chloroplasts evolved from ancient symbioses (Margulis 1981). It is therefore interesting to ask whether more modern symbionts are also evolving toward organelles, and if not, then why not? A key feature that would lead to conversion of symbionts into organelles is the reduction in symbiont genome size and transfer of vital symbiont genes to the nucleus of the host. This process appears to have occurred in some symbionts. For example, the protist *Cyanophora paradoxa* contains cyanobacterial symbionts with dramatically reduced genome size relative to free-living cyanobacteria; approximately 80 per cent of their proteins are apparently encoded by the host genome (Trench 1991). In contrast, the *Buchnera* endosymbionts have large genome sizes, even though their symbiosis with aphids is ancient, going back at least 200 million years (Moran and Baumann 1994). Thus, reduction in genome size is apparently not inevitable, and the factors that may cause symbionts to evolve in different directions are, at this point, unclear.

What factors may lead to the reduction in genome size and movement of vital genes to the nuclear genome? We can imagine several processes that could be involved, including:

(1) loss of unnecessary genes;

(2) intrahost competition among symbionts for increased replication rate;

(3) mutational degeneration (Muller's ratchet); and

(4) male-function degeneration.

As with other parasites, symbionts will lose unnecessary biochemical functions when the products are already provided by the host environment. This process probably explains the reduced genome size in rickettsia and chlamydia, obligatory intracellular parasites. However, the process does not predict transmission of vital symbiont genes to the nuclear genome, because the products are already provided by the nuclear genome.

Intrahost competition is one possible explanation for genome streamlining in incipient organelles. According to this model, if symbionts with smaller genome sizes have higher replication rates, then symbionts with deletions can increase by intrahost selection. As the defective symbiont became more abundant within hosts, this would select for compensatory mutations on the part of the host. One such mutation would be translocation of the functional gene to the nucleus, assuming the gene was appropriately expressed to rescue the defective variants. This model assumes that the product was not vital for individual symbiont survival and replication, or that the product was diffusible so that functional symbionts rescue defective symbionts in heteroplasmic hosts.

A second mechanism for genome streamlining is mutation accumulation. Because most heritable symbionts are effectively asexual, occur in small discrete populations within hosts, and undergo bottlenecks each generation, deleterious mutations are expected to accumulate via genetic drift , a process known as Muller's ratchet. During mutational degradation of particular genes, rare translocations of the functional allele (presumably present in some symbionts) to the nucleus would be selectively favoured if they enhanced symbiont performance (and therefore host survival). Similarly, increase of symbiont genes that disrupted male function (either by active selection or drift) would select for nuclear translocations of the symbiont allele that functions within males. The extent to which streamlining occurs within different symbionts, and the extent to which these different processes may be involved, is currently unknown.

1.6 Conclusion

Evolutionary interactions between inherited symbionts and hosts can be complex. In some cases, evolutionary interests of the symbiont and host are concordant, but in many cases they are not. In addition to mutualism, a variety of alternative adaptive strategies are available to heritable symbionts, including induction of cytoplasmic incompatibility, killer and Medea phenotypes, and manipulation of host sex determination. The tools provided by recent advances in molecular biology have reinvigorated studies of microbial symbionts. These fastidious microorganisms can now be identified based on gene sequence information, their genomes can be characterized, and interactions with hosts can be studied at a level not previously possible. An increasing number of studies over the past decade have demonstrated that not only are classical nutritional mutualistic symbionts very common, as has been suspected for many years, but that intracellular reproductive symbionts are also pervasive among invertebrates. It is quite possible that many more examples of inherited symbionts employing strategies of reproductive manipulation are yet to be discovered and characterized. Much work remains to be done in order to address the mechanisms by which these microorganisms are able to subvert the machinery of host reproduction to their own advantage.

Acknowledgements

The authors would like to thank Rolf Weinzierl, Ary Hoffmann, Steve Perlman, Cesar Perez-Gonzalez and Rebecca Weston for comments on the manuscript. JHW thanks Bryant McAllister and Renee Goodwin for long-distance assistance, Uzi Nur for discussions, Nico Michaels and Leo Beukeboom for an excellent sabbatical environment, and the NSF and Alexander von Humboldt Foundation for financial support. SLO thanks Thierry Rigaud, Richard Stouthamer and François Rousset for help in compiling Table 1.2, and the NIH and WHO/TDR for financial support.

2 Cytoplasmic incompatibility in insects

Ary A. Hoffmann and Michael Turelli

2.1 Introduction

Cytoplasmic incompatibility (CI) was first described in mosquitoes by Ghelelovitch (1952) and studied in detail by Laven in the 1950s (reviewed in Laven 1967*b*). Laven observed that crosses between different strains of mosquitoes sometimes failed to produce progeny, or produced progeny only when crossed in one direction (i.e. crosses of males from one strain with females from another produce progeny, but not the reciprocal cross). Laven showed that cytoplasmic factors were involved. This led to the isolation of *Wolbachia* microorganisms as the causative agent of incompatibility by Yen and others.

Following this early mosquito work, cytoplasmic incompatibility was found in several other insects, including *Nasonia*, planthoppers, raisin moths, weevils, and *Drosophila*. Soon after incompatibility was described in *Drosophila*, major advances were made in understanding both the population dynamics of incompatibility systems and the molecular evolution of incompatibility factors. Molecular markers for the agent were constructed by O'Neill *et al.* (1992) and used to investigate the nature and distribution of incompatibility factors in populations and among species. These markers also enabled phylogenetic analyses of *Wolbachia* and their hosts (Chapter 1).

In this overview, we describe the incompatibility systems found in insects, starting with Laven's work on incompatibility in mosquitoes, and including the more recent work with *Drosophila*, *Nasonia*, and other insects. We consider the mechanism of incompatibility and effects of *Wolbachia* density on incompatibility. The population dynamics of incompatibility systems are examined by looking at *Wolbachia* transmission, levels of incompatibility, sperm competition, and host fitness. We consider simple models based on these effects and compare predictions with spatial and temporal patterns of infection frequencies in nature. We also examine the association between *Wolbachia* and mitochondrial DNA (mtDNA) variation, and finally consider possible evolutionary changes in *Wolbachia* and host genomes.

2.2 Cases of cytoplasmic incompatibility due to *Wolbachia* infections

2.2.1 Mosquitoes

In the 1950s, several workers observed that crosses among strains of mosquitoes often resulted in no adult offspring or relatively few adult offspring. Crossing relationships between some European, Asian, and American strains of *Culex* (Laven 1957) are provided in Table 2.1 and indicate that complex incompatibility patterns exist. Strains do not appear to fall into discrete incompatibility groups, but instead can be characterized only by their crossing patterns with several other strains. Incompatibility between strains may be *bidirectional*, in which case no or few progeny are produced in both sets of reciprocal crosses between populations. An example from Table 2.1 is the cross between Oggelshausen and Delhi. Incompatibility may also be *unidirectional*, when males from one strain are incompatible with females from the other strain, but the reciprocal cross appears fertile. The cross between Oggelshausen and Texas provides an example of this pattern.

Several hypotheses were proposed to account for such incompatibility, including interpretations based on nuclear factors, cytoplasmic factors, and interactions between these (Smith-White and Woodhill 1954). Studies by Laven (1957) implicated a cytoplasmic factor. Laven undertook crosses with strains of *Culex* from Hamburg (Ha) and Ogglehausen (Og). The cross Ha♀ × Og♂ was compatible and produced an egg hatch of 87.3 per cent, while only around 0.2 per cent of eggs hatched in the reciprocal cross, even though mating and fertilization had taken place. There was compatibility when the F_1 females from a cross between Ha females and Og males were backcrossed to Og males. This compatibility was maintained during successive generations of backcrossing. In contrast, males from the backcross generations continued to be incompatible with Og females. These results indicated strict maternal inheritance of the factor causing incompatibility.

As well as occurring in crosses between geographically isolated populations of *C. pipiens*, incompatibility has also been found in crosses between strains from the same population. Barr (1980) found that some field egg rafts from a population of *C. pipiens* were incompatible. Magnin *et al.* (1987) observed incompatibility among 11 strains of *C. pipiens* collected from sites within 100 km of each other. Some strains collected within a few metres showed partial incompatibility (50–85 per cent of eggs in a raft hatch versus 80–99 per cent hatch from intrastrain crosses), although strong incompatibility (<25 per

Table 2.1 Incompatibillity relationships among 10 strains of *Culex* mosquitoes (after Laven 1957)

	Males									
Females	**Cal.**	**Tex.**	**Ala.**	**Geo.**	**Kan.**	**Illi.**	**Delhi**	**Kua.**	**Ham.**	**Ogg.**
California	✓	✓	✓	✓	✓	✓	✓	✓	✓	
Texas	✓	✓	✓	✓	✓	✓	✓	✓	✖	✖
Alabama			✓	✓	✓	✓		✓	✓	✖
Georgia	✓	✓	✓	✓	✓	✓	✓	✓	✓	✓
Kansas	✓	✓	✓	✓	✓	✓	✖	✓	✓	✖
Illinois	✓	✓	✓	✓	✓	✓	✓	✓	✓	
Delhi	✓	✓	✓	✖	✖	✓	✓	✓	✓	✖
Kuala Lumpur	✖	✓	✓	✓	✓	✖	✓	✓	✖	✓
Hamburg	✓	✓	✓	✓	✓	✓	✖	✖	✓	✓
Oggelshausen	✓	✓	✓	✓	✓	✖	✖	✖	✖	✓

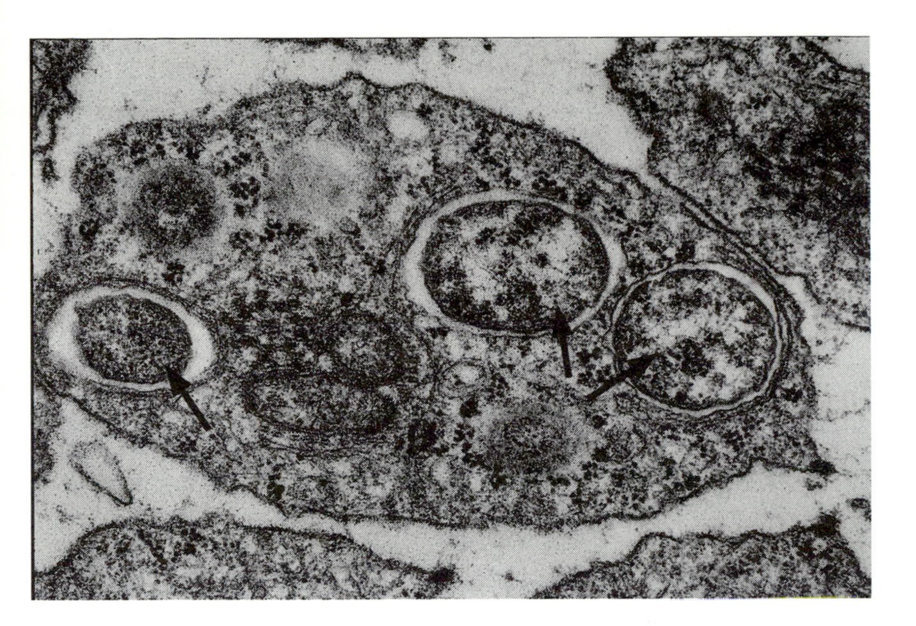

Figure 2.1 Transmission electron micrograph of *Wolbachia* (arrowed) within a developing spermatid of the moth *Ephestia cautella* (photo. by Scott ONeill).

cent hatch rate) was observed only between strains collected more than 30 m apart. As in more widespread geographic comparisons of *Culex*, the compatibility relationships among strains appeared to be complex in that strains showed distinctive patterns of incompatibility in crosses with a range of other strains. However, the role of maternally inherited microorganisms was not confirmed in all of the incompatible crosses.

Rickettsia-like microorganisms known as *Wolbachia* were first described by Hertig (1936) in the *C. pipiens* complex. Incompatibility in *C. pipiens* was first linked to *Wolbachia* infections by Yen and Barr (1973, 1974). They showed that the *Wolbachia* infection of *C. pipiens* (which they called *W. pipientis*) was widespread in the early development of embryos but eventually became restricted to pole cells. *Wolbachia* were largely restricted to germ cells of adults (Fig. 2.1). The microorganisms could be eliminated from cells by an antibiotic (tetracycline), and this procedure restored compatibility between previously incompatible strains.

Other cases of incompatibility in mosquitoes were also linked to an infective agent. In the *Aedes scutellaris* species complex, unidirectional incompatibility between populations is common. Trpis *et al.* (1981) were able to restore compatibility between two 'species', *Ae. kesseli* and *Ae. polynesiensis*, by exposing them to heat or tetracycline, suggesting that incompatibility was mediated by an infection. *Wolbachia*-like organisms were identified in *Aedes polynesiensis* by

Wright and Barr (1981). However, other factors unrelated to bacterial infection have also been associated with incompatibility in this species complex (Meek and Macdonald 1984).

The *Wolbachia* infection in mosquitoes is not always associated with strong incompatibility. For instance, Singh *et al.* (1976) found that incompatibility in *Culex* was partially lost when matings involved older males; when males were 16–17 days old, 26 per cent of egg rafts were completely (rather than partially) incompatible, compared to 90.9 per cent for 6–7-day-old males.

Recent data suggest that infections causing incompatibility may involve more than one strain of *Wolbachia* (i.e. strains are superinfected). This follows earlier work in *Nasonia* wasps (Breeuwer *et al.* 1992) showing the presence of two different *Wolbachia* ribosomal types within the same strains (see below). In mosquitoes, molecular characterization of the Houston strain of *Aedes albopictus* suggested the presence of *Wolbachia* superinfections (Sinkins *et al.* 1995*b*). These may induce patterns of incompatibility different from those shown by stocks with one infection. Sinkins *et al.* (1995*b*) hypothesized on the basis of *Drosophila* evidence (discussed below) that mosquito strains carrying an A and B infection will be unidirectionally incompatible with strains carrying either the A or B infection. In this way, stocks carrying a single infection may behave like an uninfected stock in crosses. A possible example of this phenomenon involves a strain of *Aedes albopictus* from Mauritius. Females from this strain are incompatible with males from other strains, while the reciprocal cross is compatible (Kambhampati *et al.* 1993) even though Mauritius females carry a *Wolbachia* infection. These mosquitoes may carry a single infection, resulting in unidirectional incompatibility with double-infected strains (Sinkins *et al.* 1995*b*).

The possible existence of multiple infections in mosquitoes has two implications. First, it means that testing for effects of *Wolbachia* on incompatibility should ideally be undertaken by crossing individuals to uninfected stocks. These can either be collected from natural mosquito populations or generated in the laboratory by exposure to tetracycline. Secondly, if superinfections are common, they may help to explain the complex patterns of incompatibility seen in crosses among *Culex* and *Aedes* strains. For instance, consider the case where three infections occur together in various combinations. If we assume that females carrying multiple infections (say A, B, and C) are compatible with males carrying one or more of the same infections (i.e. males carrying only A, B or C, or combinations of these infections), then the incompatibility relationships in Table 2.2 are possible. These are reminiscent of the complex patterns observed by Laven and others (Table 2.1). Another possibility is that incompatibility is controlled by different genes in different *Wolbachia* strains, leading to the same complex incompatibility patterns as superinfections. However, this possibility cannot be tested until genes causing incompatibility are identified.

Table 2.2 Putative incompatibility relationships among strains carrying one, two or three infections, based on the assumption that females are compatible with males carrying the same infection

Male infection	**Infection carried by females**						
	A	B	C	AB	AC	BC	ABC
A	✓	✖	✖	✓	✓	✖	✓
B	✖	✓	✖	✓	✖	✓	✓
C	✖	✖	✓	✖	✓	✓	✓
AB	✖	✖	✖	✓	✖	✖	✓
AC	✖	✖	✖	✖	✓	✖	✓
BC	✖	✖	✖	✖	✖	✓	✓
ABC	✖	✖	✖	✖	✖	✖	✓

✖, Incompatible; ✓, Compatible.

2.2.2 *Drosophila*

2.2.2.1 *D. simulans*

In *Drosophila*, cytoplasmic incompatibility was first described in *D. simulans* from California (Hoffmann *et al.* 1986). A strain from Watsonville showed unidirectional incompatibility when crossed to a strain from Riverside. The Riverside strain behaved as if it was infected, causing strong incompatibility (egg hatch rates <10 per cent), which declined as males used in crosses became older. Compatibility could be restored permanently by culturing larvae for a generation on medium with tetracycline. Further crosses indicated that infected stocks (designated as having incompatibility type R) were restricted to southern California (Hoffmann and Turelli 1988; Hoffmann *et al.* 1990).

Electron microscopy identified the causative agent of incompatibility in *D. simulans* as being a *Wolbachia*-like microorganism. Binnington and Hoffmann (1989) showed that *Wolbachia* were present in ovaries and testes of naturally infected *D. simulans*, whereas uninfected natural strains and *D. simulans* strains treated with antibiotics did not carry these microorganisms. Similar observations were made by Louis and Nigro (1989) in a comparison of infected and uninfected *D. simulans* strains from Italy. As in mosquitoes, mature *Drosophila* sperm did not contain *Wolbachia*.

Backcrosses indicated that the infection carried by type R flies was maternally inherited (Hoffmann and Turelli 1988). However, when infected males were mated to uninfected females (incompatibility type W), infected progeny were produced at a very low frequency, suggesting paternal inheritance at a frequency of about 1 per cent. Although laboratory females passed the infection to all their progeny, there was evidence for imperfect maternal transmission when infected females were obtained from the field (Hoffmann *et al.* 1990; Turelli *et al.* 1992; Turelli and Hoffmann 1995). We discuss these data below when considering the population dynamics of incompatibility systems.

Since the discovery of CI type R, several other incompatibility types and infections have been identified in *D. simulans* (Table 2.3). Some of these occur naturally, while others have been generated under laboratory conditions. Infected strains from Hawaii and various other Pacific and Indian Ocean islands carry *Wolbachia* that produce incompatibility type H which is bidirectionally incompatible with type R (O'Neill and Karr 1990; Montchamp-Moreau *et al.* 1991).

Montchamp-Moreau *et al.* (1991) identified bidirectional incompatibility between type R and a strain from the Seychelles, and designated the new CI type as S. Sequencing of the 16S rDNA gene indicated differences at the molecular level between the infections in type R and type S flies. In addition, Rousset *et al.* (1992*a*) showed that the S type infection was not homogeneous, but consisted of two types of *Wolbachia*, based on the presence/absence of a restriction site. Because strains may carry more than one infection, incompatibility types and infections have been distinguished in *D. simulans* (Table 2.3). The two infections carried by type S flies have been designated as *w*Ha and *w*No, while the type R infection is indicated by *w*Ri. The infection carried by type H flies is indistinguishable from *w*Ha at the molecular level. Merçot *et al.* (1995) and Rousset and Solignac (1995) have subsequently shown that some flies from the Seychelles and New Caledonia may be infected by only *w*Ha, while others carry both the *w*Ha and *w*No infections. Flies that carry only *w*Ha behave in crosses like those described from Hawaii by O'Neill and Karr (1990).

A strain carrying only the *w*No infection has recently been isolated from a Noumean strain and characterized as a new incompatibility type, designated N (Merçot *et al.* 1995). Type N males are incompatible with uninfected females, while type N females are incompatible with S males carrying both *w*Ha and *w*No as well as with CI types H and R. Type N males cause lower levels of incompatibility in crosses with uninfected females than H, S or R males. Strains that are type N have not yet been identified from natural populations.

Infections have been transferred between strains by microinjection techniques (Boyle *et al.* 1993). Rousset and De Stordeur (1994) introduced *w*Ri and *w*Ha into the genetic background of an uninfected strain. These experiments showed that the two infections behaved largely as expected on the basis of the *Wolbachia* infections rather than nuclear background. However, infection levels were lower in the experimentally infected lines compared to the donor lines. Rousset and De Stordeur (1994) suggested that this may reflect weak

Table 2.3 Incompatibility types and associated infections in *Drosophila simulans*

Incompatibility type	Infections	Incompatibility between types[a]						Comments
		R	S	H	N	A	W	
R	*w*Ri	–						Naturally occurring
S	*w*Ha, *w*No	bi	–					Naturally occurring
H	*w*Ha	bi	uni	–				Naturally occurring
N	*w*No	bi	uni	bi	–			Induced by segregation
A/M	Not named/*w*Ma	uni	uni	uni	?	–		Naturally occurring
W	–	uni	uni	uni	uni	None	–	Naturally occurring
Not named	*w*Ri, *w*Ha	uni	?	uni	?	?	uni	Generated by microinjection
Not named	*Wolbachia* from *Aedes*	bi	?	bi	?	?	uni	Generated by microinjection

[a]bi = Bidirectional, reduced egg hatch in both reciprocal crosses; uni = unidirectional, reduced egg hatch in one cross.

selection against any eggs that carry low infection levels. Consistent with this, Turelli (1994) has shown that selection on *Wolbachia* variants will tend to lower infection levels if these increase female fecundity without inducing partial incompatibility with infected males (see below).

Strains carrying double infections have been generated experimentally by introducing cytoplasm from an infected strain into a recipient strain carrying a different infection. Sinkins *et al.* (1995*b*) injected embryos from a type R strain with cytoplasm containing *w*Ha to generate a strain carrying both the *w*Ri and *w*Ha infections. The double-infected strain showed strong incompatibility with type R and type H females, while females from this strain were compatible with type R and type H males. Strains carrying both infections have not yet been identified from natural populations.

Infections that do not cause incompatibility have been isolated from natural populations of *D. simulans* (Turelli and Hoffmann 1995; Hoffmann *et al.* 1996). This sort of infection is widespread in Australian populations where it occurs at a low frequency, and has also been found in Ecuador and Florida. Females carrying the infection are incompatible with type R and type H males, and effectively behave as uninfected type W individuals. An infected strain that does not cause incompatibility has also been described from Madagascar (Rousset and Solignac 1995). The infection has been characterized at the molecular level and designated as *w*Ma; females with this infection behave like uninfected females in crosses with males of types R, H, and S. The Madagascar and Australian infections may be unrelated because they are associated with different mtDNA variants. However, it is also possible that a single infection existed prior to *D. simulans* becoming cosmopolitan; its absence from some areas may be associated with the invasion of *w*Ha or *w*Ri as predicted by theory (see below).

Finally, *Wolbachia* from *Aedes* mosquitoes have been transferred to *D. simulans* (Braig *et al.* 1994). The injected strain showed strong bidirectional incompatibility with type R and type H *D. simulans*, and unidirectional incompatibility in crosses with uninfected *D. simulans*. This suggests that the same *Wolbachia* strain may cause incompatibility in a range of nuclear backgrounds.

2.2.2.2 *D. melanogaster*

In this species, incompatibility was first described in crosses between Australian populations from Melbourne and Townsville (Hoffmann 1988). Incompatibility was relatively weak, in that 10–25 per cent of eggs failed to hatch in laboratory crosses with young males. Holden *et al.* (1993) also reported a *Wolbachia* infection in *D. melanogaster* following probing with a bacterial cell-division gene, *ftsZ*. They showed that the infection was maternally inherited, and did not cause incompatibility in crosses to uninfected flies, although weak incompatibility could not have been detected in their experimental design.

Many populations of *D. melanogaster* are polymorphic for *Wolbachia* infections. Solignac *et al.* (1994) assayed 266 lines from around the world and found an infection frequency of 34 per cent overall, and many populations contained both infected and uninfected lines. Infected individuals were obtained from all continents. The infection status of lines was confirmed by crosses, PCR amplification of 16S rRNA using *Wolbachia*-specific primers, and staining of sperm cysts with DAPI (diamidino-2-phenylindole) following a technique described by Bressac and Rousset (1993). Incompatibility levels appeared to vary widely, from apparent compatibility to hatch rates of 33 per cent, although it has not been shown conclusively that these differences represent statistically significant variation in incompatibility levels.

In Australian *D. melanogaster* populations, Hoffmann *et al.* (1994) used DAPI staining of embryos (as described in O'Neill and Karr 1990) to show that infected flies were widespread and that all Australian populations were polymorphic for this infection. Although incompatibility levels were low in a laboratory environment, they could not be increased by selection. This contrasts with the apparently high levels of variability in incompatibility evident in the survey by Solignac *et al.* (1994).

2.2.2.3 Other *Drosophila* species

Giordano *et al.* (1995) and Rousset and Solignac (1995) have recently described *Wolbachia* infections from *D. mauritiana* and *D. sechellia* strains, although not all strains of these species were infected. In *D. mauritiana*, there was no evidence for incompatibility in crosses between infected and uninfected strains. Microinjection experiments were used to transfer the *D. mauritiana* infection to type W *D. simulans* (Giordano *et al.* 1995). The infection still failed to cause incompatibility in this host; in addition, infected females from the microinjected line showed low egg hatch rates when mated to males from infected type H and type R *D. simulans* stocks. This suggests that the lack of incompatibility (or weak incompatibility) is a consequence of the *Wolbachia* strain rather than the host nuclear background. Giordano *et al.* (1995) also generated a *D. mauritiana* stock infected with *w*Ri from *D. simulans*. This strain was incompatible with naturally infected *D. mauritiana* as well as uninfected strains, indicating that *Wolbachia*-induced incompatibility was expressed in the *D. mauritiana* nuclear background. The *Wolbachia* infection carried by *D. mauritiana* was indistinguishable at the molecular level from the *w*Ma infection of *D. simulans* (Rousset and Solignac 1995). The *w*Ma infection of *D. simulans* is associated with an mtDNA variant that is closely related to the *D. mauritiana* mtDNA type. This suggests that both the infection and mtDNA were present prior to the speciation event separating these closely related species, or that the infection was spread by introgression.

In *D. sechellia*, matings between infected males and uninfected females resulted in fairly strong incompatibility (27 per cent hatch rate) when flies were 4–5 days old (Giordano *et al.* 1995). This incompatibility was no longer evident in crosses with older males (19–20 days old), consistent with data for *D. simulans* type R (Hoffmann *et al.* 1990; Turelli and Hoffmann 1995). Rousset and Solignac (1995) examined molecular variation in *Wolbachia* infections from *D. sechellia*. Three strains possessed one infection (*w*Sh), and two of these also carried a second infection (wSn). Sequence data indicated that the *w*Sh *Wolbachia* infection from D. sechellia is closely associated with the *w*Ha infection from *D. simulans*, while the *w*Sn infection is closely related to *w*No. Rousset and Solignac also considered the association between infections and mtDNA variation, and found that the *D. sechellia* mtDNA type is closely related to the *D. simulans* mtDNA type associated with the *w*Ha and *w*No infections. Both findings suggest that infections and mtDNA variants are ancestral to the speciation event separating *D. simulans* and *D. sechellia*.

Recently, Werren and Jaenike (1995) found that two (*D. recens*, *D. orientacea*) out of 10 species belonging to the closely related *quinaria*, *testacea* and *tripunctata* groups were infected with *Wolbachia*. The infection in *D. recens* was cured by exposure to tetracycline. Crosses between infected males and uninfected females produced a relatively high level of incompatibility. Bourtzis *et al.* (1996, 1994) surveyed for *Wolbachia* in 30 *Drosophila* species, and found infections in three species (*D. sechellia*, *D. auraria* and *D. ananassae*). Incompatibility levels were relatively high in *D. sechellia* and *D. auraria* but weak in *D. ananassae*.

2.2.3 *Tribolium*

Laboratory strains of the flour beetle, *Tribolium confusum*, show unidirectional incompatibility. This was initially detected in crosses between two strains (McGill Black, bI) and other *Tribolium* strains (Stanley 1961; Wade and Stevens 1985). Females from these two strains produced inviable eggs when mated with males from other strains. The ability of males to cause incompatibility was shown to be cytoplasmically inherited.

Wade and Stevens (1985) found that compatibility between bI and two other strains (bSM, b-Circle A) could be permanently restored by rearing beetles on medium with tetracycline. Moreover, treated bSM and b-CircleA strains became incompatible with untreated beetles from the same strains, showing that they carried an infection causing incompatibility, whereas bI was uninfected. Compatibility could also be restored at least partially by raising larvae at a high temperature (37 °C) for several days (Stevens 1989). *Wolbachia*-like organisms were found in the ovaries of *T. confusum* strains expected to carry the infection, and these could be eliminated with tetracycline (O'Neill 1989). The distribution of the infection in natural *Tribolium* populations is unknown. However, laboratory strains that have been established relatively recently tend to be infected (Wade and Stevens 1985), suggesting that the infection may be widespread in nature.

2.2.4 *Nasonia*

In the parasitic wasp, *Nasonia vitripennis*, incompatibility was first reported by Saul (1961) in two inbred laboratory stocks carrying mutant markers. Mated female wasps normally produce around 85 per cent female progeny in laboratory crosses, whereas unfertilized females produce only haploid male offspring. However, when Saul crossed mutant strains to wild-type males, one of the strains produced only male progeny and the other produced only 8 per cent females. In contrast, the sex ratio in the reciprocal cross was normal, suggesting unidirectional incompatibility.

The incompatibility factor in *Nasonia* was maternally transmitted (Saul 1961) and caused the loss of all paternal chromosomes in fertilized eggs (Ryan *et al.* 1985; Reed and Werren 1995). This means that any progeny arising in the cross are haploid, accounting for the male-biased sex ratio. Compatibility among strains can be restored by antibiotics; by injecting host larvae and pupae with tetracycline, Richardson *et al.* (1987) generated males compatible with females from a strain with which they were previously incompatible. Incompatibility characteristics can be transferred between strains. Williams *et al.* (1993) obtained homogenates from mutant pupae and injected them into pupae from a wild-type strain. This resulted in 39 per cent of the individuals from the wild-type strain acquiring the incompatibility properties of the mutant strain.

Nasonia individuals can harbour double infections. Breeuwer *et al.* (1992) first showed the presence of multiple 16S sequences in some strains of *Nasonia*, and hypothesized that this was due to multiple infections or a single *Wolbachia* strain with different 16S rDNA genes. Werren *et al.* (1995*a*) sequenced the *ftsZ* gene from several species and found suggestive evidence of multiple infections in *Nasonia* as well as many other species.

Recently, Perrot-Minnot *et al.* (1996) confirmed that some strains of *Nasonia* carry two distinct *Wolbachia* whereas others carry only a single infection. These workers took a *Nasonia* strain carrying two types of *Wolbachia ftsZ* sequences (A, B) and placed larvae into a prolonged diapause. After emergence from diapause, some individuals harboured both sequences, some harboured only the A sequence, some only a B sequence, and others were uninfected. Single-infected strains can therefore segregate from superinfected strains. Perrot-Minnot *et al.* (1996) also showed that lines carrying only A or B were bidirectionally incompatible, and that males from double-infected lines were incompatible with females carrying either of the single infections.

2.2.5 Other insects

Crossing data from several other insects suggest CI, but conclusive evidence is often lacking. Given that *Wolbachia* are present in 10–20 per cent of all insects (Werren *et al.* 1995*b*), numerous additional cases of CI are expected and will no doubt be detected in future work.

Evidence for CI has been obtained in three other insects. In the almond moth, *Ephestia cautella*, unidirectional incompatibility has been described between geographically isolated populations (Brower 1976). The incompatibility factor was shown to be maternally transmitted, and *Wolbachia* was identified from gonadal tissue by electron microscopy (Kellen *et al.* 1981). Compatibility could be restored by rearing larvae in the presence of an antibiotic, suggesting that CI resulted from an infective agent.

In the alfalfa weevil, *Hypera postica*, unidirectional incompatibility was described between populations from the eastern and western parts of the USA (Hsiao and Hsiao 1985). Gonads from weevils from western populations contained *Wolbachia* microorganisms, whereas those from eastern populations did not (Hsiao and Hsiao 1985). Antibodies have been produced against the *Wolbachia* infection, and a 135 kDa protein restricted to infected weevils has been identified (Leu *et al.* 1989). The functional significance of this protein is not known.

Finally, Noda (1984*a*,*b*) found that cytoplasmic factors determined incompatibility among two groups of strains of a homopteran, the small brown planthopper (*Laodelphax striatellus*). One of the strains was from the north-eastern part of Japan, while the other was from the south-western part. Incompatibility was unidirectional and partial. The progeny produced in crosses were of hybrid nuclear origin, indicating that fertilization had taken place in incompatible crosses (Noda 1987).

We conclude this section with a note about testing for incompatibility. In most cases of CI considered to date, incompatibility has been strong and therefore evident when egg hatch rates are examined after undertaking crosses between a few individuals or between two groups of individuals. However, this procedure is not adequate when testing for partial incompatibility or when comparing quantitative rather than qualitative differences in hatch rates between crosses. Eggs may fail to hatch for reasons unrelated to incompatibility, such as the absence of mating or nuclear genes causing egg inviability. Such factors cannot be taken into account when eggs are obtained from mass matings and comparisons between two crosses are based on hatch rates in two groups of eggs. For instance, if most eggs in a group are laid by a female that fails to mate, a low hatch rate for that group may result, leading an investigator to suspect incompatibility even when it is absent. For this reason, it is important to obtain an independent set of hatch rate values when comparing two or more crosses. For instance, eggs can be obtained from several pairs of males and females for each type of cross. Hatch rates of the crosses can then be compared statistically using rates from each male and female pair as independent data points in any analysis.

2.3 Infection levels and the mechanism of incompatibility

In *Culex pipiens*, incompatible sperm appear to enter eggs and induce the two meiotic divisions, but the sperm do not fuse successfully with the female pronuclei (Jost 1971). This results in a haploid egg, which may show limited development before the embryo dies unless diploidy is restored by rare parthenogenesis. There is very little embryo development observed in incompatible crosses with *Drosophila* (O'Neill and Karr 1990) and almond moths (Kellen *et al.* 1981), suggesting that *Wolbachia* effects generally occur early in development.

In the haplodiploid species, *Nasonia vitripennis*, *Wolbachia* disrupt the condensation of chromosomes from the male parent at the first mitotic division of an embryo (Ryan *et al.* 1985; Reed and Werren 1995). As a consequence, these chromosomes become a tangled mass and fragment, although genetic data indicate that genes from male parents may occasionally be transmitted (Ryan *et al.* 1985). This rare transmission probably occurs because chromosomal fragments may sometimes become associated with dividing nuclei carrying chromosomes from the female parent (Beukeboom *et al.* 1993; Reed and Werren 1995). Because cells from incompatible crosses carry only one complement of chromosomes, progeny are male. Although the phenotypic outcome of cytoplasmic incompatibility is quite different in haplodiploid species when compared to other insects such as *Drosophila* and mosquitoes, there is no a priori reason to believe that the underlying *Wolbachia*-mediated mechanism differs.

The reason why chromosomes fail to condense and embryos die is unclear. Because *Wolbachia* are not present in mature sperm (Yen and Barr 1973; Binnington and Hoffmann 1989), they must exert their effect on immature sperm which is then carried over to the egg during fertilization. This 'imprinting' of sperm might occur through the incorporation of *Wolbachia*-secreted products during sperm maturation or else through the modification or removal of normal constituents of the sperm cell during the maturation process. The presence of the same strain of *Wolbachia* in females confers compatibility, suggesting that *Wolbachia* produce factors in the egg that directly or indirectly 'rescue' the imprinted sperm.

Data from bidirectional strains and double-infected strains suggest that there has to be a close correspondence between the male 'imprinting' and female 'rescuing' components of incompatibility. For instance, the double-infected

D. simulans strain carrying *w*Ha and *w*Ri generated by Sinkins *et al.* (1995*b*) was incompatible with both R and H females, while double-infected females were compatible with males from both strains. This suggests that the *w*Ri and *w*Ha infections are associated with imprinting that can be rescued only by females carrying the same infections. As such it is difficult to construct a model of a CI 'imprinting' mechanism that relies on the removal or modification of a particular host component of the sperm cell, unless such a model can generate the variability of 'imprinting' and the specificity of 'rescue' required to explain the observed phenomenology.

The expression of incompatibility is known to be influenced by the density of bacteria in cells. In *D. simulans*, electron microscopy indicated that *Wolbachia* density was lower in tissue from older males (Binnington and Hoffmann 1989), consistent with the lower levels of incompatibility in older males (Hoffmann *et al.* 1986). Using DAPI staining, Bressac and Rousset (1993) found that the number of infected cysts in *D. simulans* decreased with male age. In *D. melanogaster*, strain differences in incompatibility levels have been related to the density of bacteria; Solignac *et al.* (1994) found a correlation between bacterial density in sperm cysts and levels of incompatibility based on 24 *D. melanogaster* strains from around the world.

In *Nasonia*, there is also evidence that bacterial density influences incompatibility levels. Breeuwer and Werren (1993) found that *Wolbachia* density was higher in a strain of *Nasonia* causing strong incompatibility than in a strain causing only partial incompatibility. However, these strain differences may be confounded by the fact that one strain is singly infected whereas the other is superinfected (Werren, personal communication). Breeuwer and Werren (1993) also found that when infected females were fed a tetracycline–sucrose solution, there was a decrease in microbe density. This was associated with a decrease in compatibility (i.e. increasing number of males produced) when treated females were mated to infected males. They showed that there was a lag phase associated with this change in CI; whereas bacterial density fell sharply after 1–2 days of antibiotic treatment, females did not show incompatibility with infected males until after 3–4 days. This suggests that bacteria determine the compatibility status of eggs as they are maturing.

Breeuwer and Werren (1993) proposed a 'bacterial dosage' model based on these and other findings. They suggested that unidirectional incompatibility results from the relative dose of bacteria in males versus females. Incompatibility will be strong when the dose in the males is high relative to the dose in the females, and become weaker as the female dose increases relative to the male dose. As indicated above, dosage alone is insufficient to account for all aspects of CI, such as bidirectional incompatibility and superinfection.

Although there appears to be an association between CI and *Wolbachia* density, this can break down in comparisons of different infections from the same or different species. In *D. simulans*, bacterial density in type A embryos is lower than in type R embryos, consistent with the fact that type A is not associated with incompatibility. However, densities in type A embryos and those from infected *D. melanogaster* are similar (Hoffmann *et al.* 1996), even though only the *D. melanogaster* infection causes some incompatibility. In *D. mauritiana*, the *Wolbachia* infection does not cause incompatibility; however, bacterial densities in embryos are higher than those in a type H *D. simulans* strain that causes strong incompatibility (Giordano *et al.* 1995). Finally, Bourtzis *et al.* (1996) found that bacterial densities in *D. melanogaster*, *D. ananassae*, *D. sechellia* and *D. auraria* were similar even though incompatibility levels differed markedly among these species. The incompatibility properties of *Wolbachia* strains may therefore be unrelated to density, at least when judged by bacterial density found in embryos. It would appear that each *Wolbachia* strain has a unique relationship between the density with which it is maintained and the subsequent strength of the CI generated.

2.4 Factors affecting the population dynamics of incompatibility

A number of factors may influence changes in the frequency of *Wolbachia* infections in natural populations, apart from their effects on CI. We briefly consider these factors before discussing how they are likely to influence infection frequencies.

2.4.1 Fitness effects on host

The effects of the *D. simulans* *w*Ri infection on host fitness have been studied intensively under laboratory conditions. Using buckets with banana pulp as population cages, Hoffmann and Turelli (1988) started populations with infection frequencies of 50 and 90 per cent. At both frequencies, they found fewer infected progeny than expected on the basis of random mating and estimated levels of incompatibility. This suggested a fitness disadvantage associated with the infection. Further experiments indicated that the infection did not influence mating or remating ability in large population cages, and that mating was random with respect to incompatibility type (Hoffmann and Turelli 1988; Hoffmann *et al.* 1990). However, there was a difference in productivity between infected and uninfected stocks, which was due to a relatively lower fecundity of infected females (Hoffmann *et al.* 1990; Nigro and Prout 1990). There was no evidence of fitness effects at other life cycle stages, at least under laboratory conditions (Hoffmann *et al.* 1990). It is not clear whether deleterious effects occur under field conditions, because infected females from the field did not lay fewer eggs than uninfected field females

(Turelli and Hoffmann 1995). Perhaps fecundity is affected only if infection levels are high, as is likely in laboratory-reared *D. simulans*

Wolbachia are not known to influence fitness in other *Drosophila* species. The infection in *D. melanogaster* has no effect on traits related to fitness, at least under optimal conditions in the laboratory (Hoffmann *et al.* 1994). The *D. mauritiana* infection (Giordano *et al.* 1995), type A infection from *D. simulans* (Hoffmann *et al.* 1996) and *Wolbachia* infections in *D. ananassae* and *D. sechellia* (Bourtzis *et al.* 1996) do not influence fecundity under laboratory conditions. Nevertheless, it seems likely that some *Wolbachia* infections in *Drosophila* do influence the fitness of their hosts. For instance, theory suggests that *Wolbachia* infections would probably not persist in *D. melanogaster* in the absence of *beneficial* host effects in this species (see below), unless the A infection is associated with a high incidence of horizontal transmission. The latter seems unlikely because the frequency of the A infection does not increase in population cages where infected and uninfected flies are held in close proximity (Hoffmann, unpublished).

Wolbachia may cause deleterious effects on *Tribolium* hosts. Stevens and Wade (1990) found that an infected strain had lower productivity than an uninfected strain. More strains need to be compared to verify this result, and data are also needed on infected beetles from natural populations.

2.4.2 Sperm competition

Potential effects of *Wolbachia* on sperm competition were first tested by Hoffmann *et al.* (1990) for type R *D. simulans*. By crossing infected females carrying a recessive mutant to wild-type males, followed by a second mating with mutant males, they directly tested the effects of the infection on the ability of the second males sperm to displace the first males sperm. These experiments showed that *Wolbachia* did not influence sperm competition. However, *D. simulans* females do not remate readily until they have utilized some sperm from the first male. This means that sperm competition could not be tested when similar-sized ejaculates from two males were present and sperm competition is likely to be most intense. Experiments with a *D. melanogaster* mutant have also failed to find an effect of *Wolbachia* on sperm competition (Hoffmann, unpublished).

In contrast, Wade and Chang (1995) provided suggestive evidence that *Wolbachia* influenced sperm competition in *Tribolium confusum*. This species remates more readily than *D. simulans*, so sperm competition may be more intense than in *D. simulans*. Wade and Chang held infected females with different numbers of infected and uninfected males. This resulted in a reduction in progeny number that was greater than expected on the basis of random mating. Because *Tribolium* were shown to mate randomly in a separate experiment, Wade and Chang (1995) suggested that *Wolbachia*-infected sperm were at an advantage in competition with uninfected sperm. If this is the case, incompatibility could be enhanced in uninfected females that have remated, increasing the rate at which the infection is expected to spread in a population (see below). However, conclusive evidence of sperm competition in this species remains to be collected.

2.4.3 Field incompatibility levels

Field incompatibility estimates have been obtained only for the *w*Ri infection of *D. simulans* (Hoffmann *et al.* 1990; Turelli and Hoffmann 1995) and for the *D. melanogaster* infection in Australia (Hoffmann, unpublished), although incompatible egg rafts have been collected from field mosquito populations (e.g. Barr 1980). For the *w*Ri infection of *D. simulans*, egg hatch rates obtained with field-collected females and field males mated to laboratory females have shown that incompatibility levels in the field are variable and generally lower than those observed in laboratory experiments with young males. Hatch rates are in the order of 30–70 per cent of those from compatible crosses. Differences between laboratory and field incompatibility levels are also evident in *D. melanogaster*; whereas laboratory experiments indicate a reduction in hatch rate of 10–30 per cent (Hoffmann 1988), no incompatibility has been detected in three collections of *D. melanogaster* from natural populations.

Incompatibility levels may be lower in the field than the laboratory for several reasons. Matings to older males can result in low levels of incompatibility, as previously mentioned. Reduced levels of incompatibility could result from temperature effects because high temperatures suppress incompatibility (Hoffmann *et al.* 1986; Stevens 1989). Another possibility is that a high larval density reduces CI. Incompatibility levels were higher when a double-infected line of *D. simulans* was raised at a low density compared to a high density (Sinkins *et al.* 1995*b*), although it is not known if density effects per se are involved rather than nutrition. Recent experiments have shown that different *Wolbachia* strains respond differently to these conditions (O'Neill, personal communication). While *w*Ha is extremely sensitive, crowding appears to have little effect on the strength of CI in type R, suggesting that this factor does not contribute to the lower field incompatibility levels in *w*Ri *D. simulans*.

Finally, naturally occurring antibiotics may suppress levels of CI or eliminate *Wolbachia* entirely. Stevens and Wicklow (1992) reared infected *Tribolium* on flour produced from wheat grain containing *Streptomyces* or *Penicillium* moulds. One strain of *Streptomyces* produced a high level of curing in both males and females, whereas exposure to several other *Streptomyces* strains did not. The frequency of curing after exposure to *Penicillium* was low; only a few cured females were obtained, whereas there were no cured males. These types of moulds may be encountered by *Tribolium* and other insects showing incompatibility, but there are no field data documenting effects of naturally occurring antibiotics. Although *Drosophila* are likely to encounter moulds in their natural environments, there is no evidence that flies raised on mouldy fruit have lower levels of incompatibility than those raised on laboratory media (Turelli and Hoffmann 1995). In addition, the frequency of uninfected *D. simulans* in predominantly infected natural populations can be adequately understood by using estimated values for incompatibility levels and imperfect maternal transmission and ignoring the possibility that larvae are cured by direct exposure to antibiotics (Turelli and Hoffmann 1995).

In *D. simulans*, type R males reared under field conditions caused high levels of incompatibility comparable to those associated with laboratory-reared males (Turelli and Hoffmann 1995). Moreover, levels of incompatibility caused by these flies changed with age at a rate comparable to that found in laboratory-reared males held in seminatural field cages. This suggests that male age may be the predominant factor contributing to lower incompatibility levels under field conditions. These findings also suggest that older males obtain most of the matings in *D. simulans* field populations. To obtain levels of incompatibility observed in the field, females would have to mate with males that are 2–3 weeks old, on average (Turelli and Hoffmann 1995).

2.4.4 Segregation/maternal transmission fidelity

In *Culex* mosquitoes, both French (1970) and Subbarao *et al.* (1977*b*) found that males with different compatibility properties could occasionally be produced from the same egg raft. For instance, Subbarao *et al.* (1977*b*) found that 2 out of 50 egg rafts from a culture that was initially infected produced mosquitoes that behaved differently from those of the parental culture. Males from one of these rafts no longer showed incompatibility with females from another culture (De), unlike males from the parental colony. Males from the other egg raft behaved inconsistently; some were compatible with De females, while others were incompatible. Subbarao *et al.* (1977*b*) isolated sublines from this egg raft, and showed that some sublines behaved like strains lacking the infection, while others behaved as if they were carrying a different infection. This suggests that segregation for incompatibility infections can occur; occasionally, infected females may produce progeny lacking the infection, or females that carry two infections may produce progeny carrying only one. However, most mosquito lines [48 out of 50 in the case of Subbarao *et al.* (1977*b*)] do not show segregation and transmit their incompatibility type with complete fidelity.

Segregation has also been described in type R *D. simulans*. In the lab, all offspring from infected mothers carry the infection, suggesting that segregation is rare or absent in this environment (Hoffmann and Turelli 1988). However, this is not the case in field populations of *D. simulans*. When field females are brought into the lab, some produce offspring that are uninfected. We initially deduced that these progeny were uninfected from their behaviour in crosses (Hoffmann *et al.* 1990). Later experiments using *Wolbachia*-specific primers (O'Neill *et al.* 1992) confirmed the absence of *Wolbachia* from such progeny (Turelli and Hoffmann 1995). In type R *D. simulans*, there was variation in the rates at which field females produced uninfected progeny. Some females failed to show segregation in their progeny, whereas around 50 per cent of progeny from other females were uninfected (Turelli and Hoffmann 1995). On average, 3–5 per cent of the progeny from infected field females were uninfected, although this rate may vary between collections.

Segregation in *D. simulans* can also occur in lines carrying more than one infection. As mentioned earlier, Merçot *et al.* (1995) isolated a *D. simulans* line carrying only the *w*No infection from a double-infected line following segregation. Sinkins *et al.* (1995b) also obtained evidence for such segregation in lines experimentally infected by both *w*Ri and *w*Ha. In *Nasonia*, Perrot-Minnot *et al.* (1996) showed that single-infected lines could be obtained from double-infected lines following larval diapause.

There are several reasons why segregation levels may be higher in the field than the laboratory. Factors such as temperature, natural antibiotics, diapause and density that influence levels of incompatibility could also affect maternal transmission. For instance, Sinkins *et al.* (1995*b*) noted that loss of an infection from a double-infected strain was higher when larvae were raised in crowded conditions than in uncrowded conditions. It seems likely that such factors reduce the density of bacteria, resulting in segregation, because not all host eggs receive *Wolbachia*. Evidence for this hypothesis comes from crosses between females reared in the field and young laboratory males that induce a high level of incompatibility (Turelli and Hoffmann 1995). Females that had a relatively high segregation rate also showed some incompatibility with these males, suggesting that segregation and incompatibility are linked via bacterial density.

Field segregation estimates have been obtained for two other *Wolbachia* infections. In type A *D. simulans*, maternal transmission appeared to be perfect or near-perfect (Hoffmann *et al.* 1996), although this conclusion is based on only a single field sample. In *D. melanogaster*, two field estimates indicated segregation rates of 2–5 per cent, similar to those observed for *D. simulans* type R. These results also suggest that, as in the case of incompatibility, there is no association between segregation and bacterial density in comparisons between different species or different infections. *Drosophila melanogaster* harbour a lower *Wolbachia* density than *D. simulans* type R even though they show the same transmission efficiency. The type A infection of *D. simulans* occurs at a low density in embryos compared to *w*Ri (Hoffmann *et al.* 1996), even though the latter shows a higher level of segregation.

2.5 Theory

The first theoretical analysis of CI population dynamics was presented by Caspari and Watson in 1959. Since then several papers have advanced the theory by incorporating newly discovered phenomena. We will provide an overview, and slight extension, of these theoretical analyses of CI dynamics within and among populations, emphasizing the factors discussed above that affect frequency dynamics in natural and/or laboratory populations. For consistency, we will use the parameterizations found in our papers. After reviewing the theory, we will discuss how parameter estimates from natural populations have been used to predict frequency changes and equilibria in nature.

2.5.1 Intrapopulation dynamics

2.5.1.1 Incompatibility with fecundity effects

Caspari and Watson (1959) determined how the frequency of an infection would change in a population if infected females suffered a fecundity loss. They assumed perfect maternal transmission of the incompatibility-inducing infection. In the absence of fecundity differences between infected (I) and uninfected (U) females, the frequency of I will increase each generation, because I females will produce the same number of I progeny irrespective of their mates, whereas U females will produce a reduced number of U progeny if they mate with I males. Caspari and Watson (1959) showed, however, that if I females are less fecund, an unstable equilibrium frequency exists above which the infection frequency increases to fixation, but below which it decreases to zero.

To quantify this interaction, Caspari and Watson assumed discrete generations and random mating between I and U individuals. Let p_t denote the frequency of I adults (assumed to be the same for males and females) in generation t, let F denote the relative fecundity of I females, and let H denote the relative hatch rates from incompatible (U♀ × I♂) versus compatible crosses (assuming that crosses I♀ × U♂, U × U and I × I produce equal hatch rates). Let $s_h = 1 - H$ and $s_f = 1 - F$. Assuming that $H \leqslant 1$ and $F \leqslant 1$, s_h measures selection favouring I based on incompatibility and s_f measures selection against I based on its reduced fecundity. Because the incompatibility advantage is frequency dependent, these two forces can produce an unstable polymorphic equilibrium. Using different notation, Caspari and Watson (1959) showed that p_t changes according to:

$$p_{t+1} = \frac{p_t(1 - s_f)}{1 - s_f p_t - s_h p_t(1 - p_t)} \tag{2.1}$$

Thus, the equilibrium frequencies for I are $p = 0$, $p = 1$ and $p = s_f/s_h$. Assuming that I females suffer a fecundity disadvantage ($s_f > 0$), loss of the infection ($p = 0$) is always a locally stable equilibrium. If $s_h \leqslant s_f$, so that the fecundity cost exceeds the incompatibility advantage, $p = 0$ is globally stable, and the infection will be eliminated from a polymorphic population. If $s_h > s_f > 0$, then $p = 0$ and $p = 1$ are both locally stable; and the ultimate frequency of I is determined by whether its initial frequency is above or below the unstable equilibrium, $p = s_f/s_h$ (Caspari and Watson 1959).

A central feature of this model is that it does not predict the persistence of polymorphic populations. This is always true if perfect maternal transmission and cytoplasmic incompatibility occur. Rousset *et al.* (1991) proved that there are no stable polymorphic equilibria for any number of infection types displaying any pattern of incompatibility and effect on fecundity. In particular, their result implies that neither single- and double-infected lines nor bidirectionally incompatible types can stably coexist in an isolated population if maternal transmission is perfect.

2.5.1.2 Imperfect maternal transmission and larval curing

Fine (1978) recognized that the simplest way to explain the stable coexistence of both infected and uninfected individuals in a natural population is to assume that infected females produce some uninfected progeny, i.e. that there is 'segregation' or imperfect maternal transmission. Let μ denote the fraction of uninfected ova produced by I females. Fine (1978) assumed that the infection affects equally the 'fertility' of both males and females. However, no data support the assumption that I males obtain fewer mates. Hence, Hoffmann *et al.* (1990) generalized Caspari and Watson's (1959) model (2.1), in which only female fecundity is affected by I, to include imperfect maternal transmission. Assuming that uninfected ova, whether they come from infected or uninfected mothers, are equally susceptible to incompatibility with sperm from infected males, equation (2.1) becomes:

$$p_{t+1} = \frac{p_t(1-\mu)F}{1 - s_f p_t(1-p_t) - \mu s_h p_t^2 F}. \tag{2.2}$$

Loss of I ($p = 0$), is locally stable if $F(1-\mu) < 1$. Hence, CI-inducing infections will not spread from very low frequencies in an isolated population if they either lower fecundity or are not always maternally transmitted. There are two feasible polymorphic equilibria, denoted p_s and p_u, if

$$1 > s_h > \frac{s_f}{1 - 2\mu F} \quad \text{and} \tag{2.3a}$$

$$0 < \mu < \frac{1}{2}\left(1 - \sqrt{\frac{H(s_{\mathrm{h}} - s_{\mathrm{f}}^2)}{s_{\mathrm{h}} F^2}}\right), \tag{2.3b}$$

where, as above, $F = 1 - s_{\mathrm{f}}$ and $H = 1 - s_{\mathrm{h}}$ (Turelli 1994). (Because of a typographical error, the term F^2 in condition (2.3b) mistakenly appears as F in Turelli (1994) and Turelli and Hoffmann (1995). Constraint (2.3b) results from requiring that the square-root term in equation (2.4), below, be positive. It can be re-expressed as a constraint on s_{f} or s_{h}.) The higher equilibrium,

$$p_{\mathrm{s}} = \frac{s_{\mathrm{f}} + s_{\mathrm{h}} + \sqrt{(s_{\mathrm{f}} + s_{\mathrm{h}})^2 - 4(s_{\mathrm{f}} + \mu F)s_{\mathrm{h}}(1 - \mu F)}}{2s_{\mathrm{h}}(1 - \mu F)} > \frac{1}{2}, \tag{2.4}$$

is stable. The lower equilibrium, p_{u}, obtained by changing the sign of the square root in equation (2.4), is unstable. (As expected, these become $p_{\mathrm{s}} = 1$ and $p_{\mathrm{u}} = s_{\mathrm{f}}/s_{\mathrm{h}}$ if $\mu = 0$.) The infection will be lost from an isolated population if its initial frequency is below p_{u}. Thus, if $F(1-\mu) < 1$, sampling drift is required for the infection to become established locally.

As discussed below, this simple model adequately describes natural *D. simulans* populations in which *w*Ri is near fixation. For completeness, we will briefly describe two generalizations that may be useful for other populations. First, one can assume that both infected and uninfected ova from infected females are compatible with sperm from infected males. Secondly, one can allow for direct curing of infected larvae, for instance, by exposure to high temperatures, crowding or other stressful conditions, or naturally occurring antibiotics. Both assumptions produce identical recursions (see Eq. 3 of Turelli *et al.* 1992). If equal values are used for the parameters of these models that are analogous to μ in equation (2.2), the resulting equilibrium infection frequency among adults is lower than that obtained from equation (2.4), which assumes that uninfected ova from infected mothers are susceptible to incompatibility and hence are partially eliminated.

2.5.1.3 Paternal transmission and the joint dynamics of mtDNA variants and CI types

Although paternal transmission of *Wolbachia* seems to be very rare or absent in natural populations of *D. simulans* (Turelli *et al.* 1992; Turelli and Hoffmann 1995), it has been demonstrated both directly (Hoffmann *et al.* 1990) and indirectly (Nigro and Prout 1990) in laboratory populations. Turelli *et al.* (1992) generalized equation (2.2) to include male transmission. When fewer than 1 per cent of the offspring from U♀ × I♂ matings are infected, paternal transmission has relatively little effect on the frequency dynamics of the infection. However, even such rare paternal transmission has a significant impact on the joint frequencies of mtDNA and CI types.

If transmission is purely maternal and horizontal transmission between taxa is an essentially unique event, one expects that if the infection spreads, so will the mtDNA haplotype that was originally infected. As demonstrated by Turelli *et al.* (1992), when maternal transmission is imperfect, both infected and uninfected individuals will carry the same mtDNA haplotype after the infection has reached the equilibrium described by equation (2.4). The reason that the 'sweep' affects both infected and uninfected individuals is that with imperfect maternal transmission, as the infection spreads in a population, more and more of the uninfected individuals will have infected maternal ancestors. Hence, an increasing fraction of the uninfected individuals will carry the mtDNA haplotype originally associated with the infection. In contrast, rare paternal transmission decouples the transmission of CI types and mtDNA, so that one expects to find multiple mtDNA haplotypes among both infected and uninfected individuals, assuming the species has multiple mtDNA haplotypes prior to infection. Turelli *et al.* (1992) present recursions that describe the joint dynamics of CI types and mtDNA haplotypes, under the simplifying assumption that infected and uninfected individuals are initially monomorphic for alternative mtDNA haplotypes (this is equivalent to assuming spatial variation in mtDNA haplotypes prior to infection). These analyses are discussed further below.

2.5.1.4 Dynamics of double infections with imperfect transmission

The models above can be simply generalized to describe more complex infections, such as the double infections recently described in *D. simulans* (Rousset and Solignac 1995), *Nasonia* (Werren *et al.* 1995*a*; Perrot-Minnot *et al.* 1996) and several neotropical insects (Werren *et al.* 1995*b*). Suppose some individuals harbour a double infection, denoted I_{AB}, some carry single infections (I_A or I_B), and some are uninfected (denoted $I_\varnothing$ or U). We will assume that the singly infected and uninfected individuals arise from imperfect maternal transmission. Let $\mu_{AB,A}$ ($\mu_{AB,B}$) denote the frequency of I_A (I_B) ova produced by I_{AB} females; let $\mu_{AB,\varnothing}$ denote the frequency of U ova produced by I_{AB} females; and let μ_A (μ_B) denote the frequency of U ova produced by I_A (I_B) females. We assume that I_{AB} ova are compatible with all males, whereas I_A (I_B) ova are incompatible with sperm from I_{AB} and I_B (I_A) males, and U ova are incompatible with sperm from males carrying any infection. Let $H_{A,AB}$ denote the hatch rate of embryos produced by I_A ova fertilized by sperm from I_{AB} males, let $H_{A,B}$ denote the hatch rate of embryos produced by I_A ova fertilized by sperm from I_B males, and let $H_{\varnothing,x}$ denote the hatch rate of embryos produced by U ova fertilized by sperm from x males, with $x = I_{AB}$, I_A, or I_B. Finally, let F_{AB}, F_A, and F_B denote the fecundities of the various types of infected females relative to U females. In total, there are 15 parameters that must be estimated: seven describing incompatibilities, three describing relative fecundities and five describing maternal transmission.

Extending the biological assumptions that led to equation (2.2), the recursions for the frequencies of the various types of adults are:

$$\bar{H}\, p_{\mathrm{AB,t+1}} = F_{\mathrm{AB}}\, p_{\mathrm{AB,t}}\,(1 - \mu_{\mathrm{AB}}), \tag{2.5a}$$

$$\begin{aligned}\bar{H}\dot{p}_{\mathrm{A,t+1}} = {} & [F_{\mathrm{A}}\, p_{\mathrm{A,t}}\,(1 - \mu_{\mathrm{A}}) + F_{\mathrm{AB}}\, p_{\mathrm{AB,t}}\, \mu_{\mathrm{AB,A}}](p_{\mathrm{AB,t}}\, H_{\mathrm{A,AB}} + p_{\mathrm{A,t}} \\ & + p_{\mathrm{B,t}} H_{\mathrm{A,B}} + p_{\varnothing,\mathrm{t}}),\end{aligned} \tag{2.5b}$$

$$\begin{aligned}\bar{H}\dot{p}_{\mathrm{B,t+1}} = {} & [F_{\mathrm{B}}\, p_{\mathrm{B,t}}\,(1 - \mu_{\mathrm{B}}) + F_{\mathrm{AB}}\, p_{\mathrm{AB,t}}\, \mu_{\mathrm{AB,B}}](p_{\mathrm{AB,t}}\, H_{\mathrm{B,AB}} \\ & + p_{\mathrm{A,t}}\, H_{\mathrm{B,A}} + p_{\mathrm{B,t}} + p_{\varnothing,\mathrm{t}}),\end{aligned} \tag{2.5c}$$

and

$$\begin{aligned}\bar{H}\dot{p}_{\varnothing,\mathrm{t+1}} = {} & [p_{\varnothing,\mathrm{t}} + F_{\mathrm{A}}\, p_{\mathrm{A,t}}\, \mu_{\mathrm{A}} + F_{\mathrm{B}}\, p_{\mathrm{B,t}}\, \mu_{\mathrm{B}}) + F_{\mathrm{AB}}\, p_{\mathrm{AB,t}}\, \mu_{\mathrm{AB},\varnothing}] \\ & \times (p_{\mathrm{AB,t}}\, H_{\varnothing,\mathrm{AB}} + p_{\mathrm{A,t}}\, H_{\varnothing,\mathrm{A}} + p_{\mathrm{B,t}}\, H_{\varnothing \mathrm{A}} + p_{\varnothing,\mathrm{t}}),\end{aligned} \tag{2.5d}$$

where $\mu_{\mathrm{AB}} = \mu_{\mathrm{AB,A}} + \mu_{\mathrm{AB,B}} + \mu_{\mathrm{AB},\varnothing}$ and $\bar{H}$ denotes the sum of the terms on the right-hand sides of (2.5a–d). Each of these expressions has a simple interpretation as the product of two terms, the first proportional to the fraction of ova of each type, the second proportional to the fraction of fertilized ova of each type that will hatch. (In equation (2.5a) this second term is 1, because doubly infected ova are compatible with all sperm.) Given that the sum of the four CI-type frequencies is 1, there are only three independent variables. An alternative parameterization is to consider $\pi_{\mathrm{A}} = p_{\mathrm{A}} + p_{\mathrm{AB}}$, the frequency of individuals that carry infection $\mathrm{I_A}$ (with or without $\mathrm{I_B}$), $\pi_{\mathrm{B}} = p_{\mathrm{B}} + p_{\mathrm{AB}}$, the frequency of individuals that carry infection $\mathrm{I_B}$, and $D = p_{\mathrm{AB}} - \pi_{\mathrm{A}}\pi_{\mathrm{B}}$, which measures the degree to which the two infections are statistically associated. If being infected with $\mathrm{I_A}$ does not influence the probability of being infected with $\mathrm{I_B}$, $D = 0$. D is precisely analogous to 'linkage disequilibrium' in two-locus population genetics theory (Hartl and Clark 1989).

A complete analysis of these recursions is beyond the scope of this chapter. However, some insight into the dynamics and equilibria of double infections can be obtained by considering a simple case in which each infection behaves 'independently' in the following sense. Suppose that:

(1) $F_{\mathrm{AB}} = F_{\mathrm{A}}F_{\mathrm{B}}$, corresponding to multiplicative effects on fecundity;

(2) $\mu_{\mathrm{AB,A}} = \mu_{\mathrm{B}}(1-\mu_{\mathrm{A}})$, $\mu_{\mathrm{AB,B}} = \mu_{\mathrm{A}}(1-\mu_{\mathrm{B}})$ and $\mu_{\mathrm{AB},\varnothing} = \mu_{\mathrm{A}}\mu_{\mathrm{B}}$, corresponding to independent loss of each infection; and

(3) $H_{\mathrm{A,AB}} = H_{\mathrm{A,B}} = H_{\varnothing,\mathrm{B}}$, $H_{\mathrm{B,AB}} = H_{\mathrm{B,A}} = H_{\varnothing,\mathrm{A}}$, and $H_{\varnothing,\mathrm{AB}} = H_{\varnothing,\mathrm{A}}H_{\varnothing,\mathrm{B}}$, corresponding to each infection acting as an independent source of mortality in incompatible fertilizations.

Under these 'independence' assumptions, it is reasonable to conjecture that the frequency of the double infection will be closely approximated by the products of the individual infection frequencies (i.e. $D \approx 0$), and that the frequency of each individual infection can be reasonably approximated by equation (2.2). This conjecture is supported by several numerical examples (see below), and it provides a simple way to begin to understand the population biology of multiple infections.

For instance, under what circumstances will a doubly infected strain successfully invade a population near equilibrium for one of the two infections? The 'independent infections' approximation suggests that I_{AB} will spread through a population near equilibrium for I_A if, and only if, the initial frequency of I_{AB} is sufficiently high that the frequency of I_B exceeds the unstable equilibrium value that would block the spread of I_B in an uninfected population. This effect is illustrated in Fig. 2.2a, which shows that this simple invasion criterion accurately approximates the full dynamics described by equation (2.5). For the parameter values used ($\mu_A = 0.04$, $\mu_B = 0.04$, $F_A = 0.8$, $F_B = 0.9$, $H_{\varnothing,A} = 0.4$, $H_{\varnothing,B} = 0.5$), the unstable and stable equilibria for each infection type are (0.42, 0.96) for A and (0.33, 0.93) for B, respectively. As seen in Fig. 2.2a, the double infection successfully invades only if the initial value of π_B exceeds 0.31, very close to the threshold 0.33 predicted from a one-infection analysis. Of course, double infections need not act independently. Nevertheless, the qualitative conclusion—that double infections are likely to successfully invade singly infected populations only when the initial frequency is sufficiently high—is certain to be robust if maternal transmission is imperfect and/or infections lower fecundity.

2.5.1.5 Infections that do not cause CI

Although our emphasis is on *Wolbachia* infections that cause CI, the type A infections of *D. simulans* (Hoffmann *et al.* 1996) and the *Wolbachia* infection in *D. mauritiana* (Giordano *et al.* 1995) among others do not seem to lower hatch rates when infected males are mated to uninfected females. When $H = 1$, recursion (2.2) simplifies considerably. A stable equilibrium will be maintained by fecundity effects and imperfect transmission if, and only if, $F(1-\mu) > 1$ (which makes $p = 0$ an unstable equilibrium) and $\mu > 0$ (which makes $p = 1$ unstable). This requires that these *Wolbachia* infections *increase* the fecundity of infected females, but that the infection not be maternally transmitted to all progeny. It is also possible that these infections are associated with a high rate of horizontal transmission, although there is, as yet, no evidence for this conjecture. When both conditions are met, the globally stable equilibrium infection frequency is

$$p_s = 1 - \frac{\mu F}{F - 1}. \tag{2.6}$$

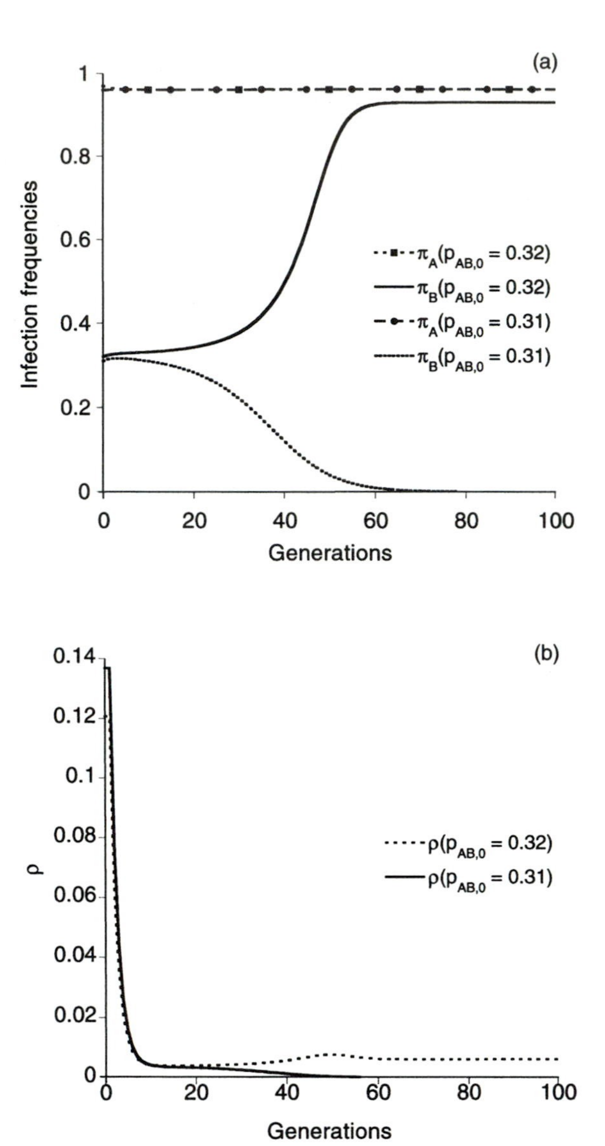

Figure 2.2 Dynamics of a double infection (I_{AB}) introduced into a population initially infected only with I_A. (a) The dynamics of the frequencies of individuals infected with I_A ($\pi_A = p_{AB} + p_A$) and I_B. (b) The association between the infections, quantified by the normalized index $\rho = D/\sqrt{\pi_A(1-\pi_A)\pi_B(1-\pi_B)}$. As expected for 'independently' acting infections, *r* quickly approaches 0 and remains near it. See the text for further explanation.

This is precisely analogous to mutation-selection balance for haploids; and the equilibrium frequency for the uninfected type can be expressed simply as $\mu/\tilde{s}_f$, where $\tilde{s}_f$ denotes the relative fecundity disadvantage of the uninfected females (i.e. $\tilde{s}_f = (F-1)/F$).

The fecundity advantages leading to equation (2.6) are purely speculative at the moment. However, even a very small advantage—of the order of a few per cent—which would be difficult to document, would suffice to maintain 'benign' infections that are transmitted with high fidelity. If such advantages are found, the calculations leading to the stable equilibrium (2.4) must be reconsidered. Although equation (2.4) remains valid with $F > 1$, the constraints for the existence of stable and unstable polymorphisms are different. In particular, if $F(1-\mu) > 1$, the stable polymorphic equilibrium described by equation (2.4) may be approached directly from $p = 0$ without passing an unstable point.

2.5.1.6 CI effects on sperm competition

Wade and Chang (1995) provide indirect evidence that sperm from I males is preferentially utilized over sperm from U males. We will briefly consider how this might affect CI population dynamics by using Prout and Bungaard's (1977) sperm competition model to generalize equation (2.2). Assuming that females carry sperm from at most two males, Prout and Bungaard (1977) used two parameters to describe sperm displacement: M describes the frequency with which sperm from two males are simultaneously present in a female, and K_I describes the average fraction of ova inseminated by I sperm when both I and U sperm are present [this parameter averages over the two possible orderings of successive matings with I and U males—see Prout and Bungaard (1977) for a detailed description]. With only two types of males, the net result of sperm competition can be described by a single parameter that summarizes both the frequency with which sperm competition occurs and its outcome. Let $\delta_I = M(2K_I - 1)$, so that $-1 \leqslant \delta_I \leqslant 1$. The Prout and Bungaard (1977) model implies that when adult I males have frequency p, the fraction of ova inseminated by I sperm is $p + pq\delta_I$, with $q = 1-p$. The resulting generalization of equation (2.2) is:

$$p_{t+1} = \frac{p_t(1-\mu)F}{1 - s_f p_t - s_h p_t(q_t + F\mu p_t)(1 + q_t \delta_I)}, \qquad (2.7)$$

with $q_t = 1-p_t$. As before, $p = 0$ is a stable equilibrium if $F(1-\mu) < 1$. Hence, drift remains necessary to establish such an infection. Because the denominator of equation (2.7) is a cubic in p_t, it is difficult to obtain a useful mathematical description of the polymorphic equilibria. However, extensive numerical analyses suggest that recursion (2.7), like (2.2), possesses at most two feasible polymorphic equilibria, with the lower one unstable. As δ_I increases, indicating a greater sperm-competition advantage for I, the stable equilibrium analogous to (2.4) increases and the corresponding unstable equilibrium decreases.

Because the net sperm displacement advantage is proportional to $p(1-p)$, biased sperm competition can have a large effect on equilibria near 0.5 but almost no effect on equilibria near 0 or 1. Figure 2.3 plots the stable and unstable equilibria as a function of δ_I for two sets of parameters. For both sets, $\mu = 0.04$ and $s_f = 0$; in one $s_h = 0.45$, in the other $s_h = 0.16$. The first set of parameters ($s_h = 0.45$) corresponds to estimates from natural *D. simulans* populations, and sperm displacement has little effect on either equilibrium. With $s_h = 0.16$, both equilibria are nearer 0.5 with $\delta_I = 0$, and increasing the advantage for I has a much larger effect. In particular, by lowering the unstable point, sperm displacement favouring I would facilitate the establishment of the infection in an isolated population by drift.

Even when sperm displacement biases have only a small effect on equilibria, they can significantly affect the rate at which infections sweep through a population. With high levels of variation, biased sperm displacement is equivalent to increasing the level of incompatibility between infected males and uninfected females. For instance, with $\mu = 0.04$, $s_f = 0$ and $s_h = 0.45$, it takes 14 generations for p to increase from 0.3 to 0.8 when $\delta_I = 0$; but this drops to 10 generations with $\delta_I = 0.5$ and 8 generations with $\delta_I = 1.0$.

2.5.2 Interpopulation dynamics: spatial spread

Turelli and Hoffmann (1991) presented an idealized analysis of the spatial

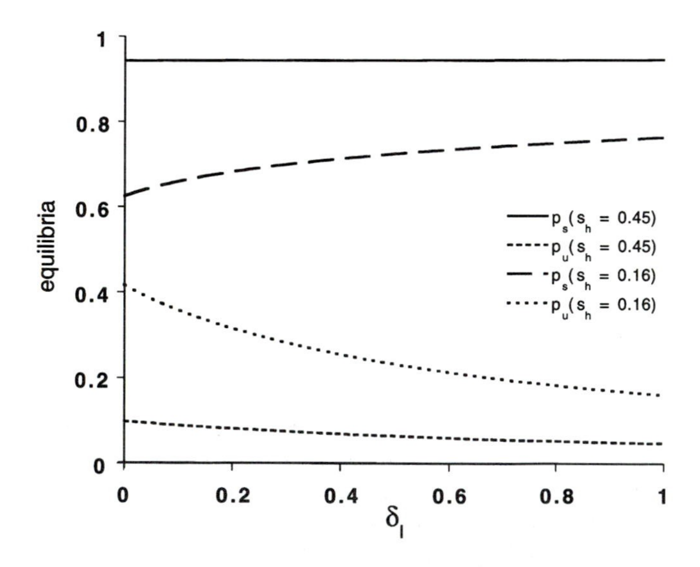

Figure 2.3 Changes in the position of the stable and unstable equilibria as a function of the level of biased sperm competition. See the text for an explanation of the symbols and the model.

spread of CI-inducing infections. We assumed that the populations were spatially distributed with a uniform population density and that individual dispersal distances follow a normal distribution. The resulting mathematical approximation was identical to Barton's (1979) generalization of Fisher's (1937) equation describing the spatial spread of an advantageous mutation. To analyse hybrid zones associated with chromosome rearrangements, Barton (1979) considered selection against heterozygotes at a diallelic locus. Like Caspari and Watson's (1959) CI model, equation (2.1), underdominant selection produces an unstable polymorphic equilibrium. Barton (1979) showed that if the unstable equilibrium frequency for allele A was less than 0.5 (implying that AA is the more fit homozygote), allele A will tend to spread spatially once it is close to fixation in a sufficiently large area. By analogy, we expect that CI-causing infections will tend to spread spatially once they become established locally, whenever the unstable equilibrium frequency is less than 0.5. Although the stable equilibrium [equation (2.4)] produced by incompatibility and imperfect maternal transmission is always greater than 0.5, the corresponding unstable equilibrium may also be greater than 0.5. Hence, if the parameters F, H, or μ in equation (2.2) change either environmentally or evolve so that the unstable equilibrium increases above 0.5, the waves of advance for CI-causing infections can be reversed, so that infection frequencies decline toward zero.

Barton (1979) produced his analysis to understand the apparently stable positions and widths of hybrid zones in nature, rather than their spatial movement. He found that although the 'tension zones' between underdominant chromosome arrangements will tend to move through homogeneous habitats with uniform population densities and dispersal parameters, these waves of advance can be stopped relatively easily by barriers to dispersal. The mathematical requirement is simply that the 'barrier' reduce dispersal. These barriers can be physical impediments such as mountains, rivers, or patches of unsuitable habitat that block dispersal. In principle, these could also be produced by an exceedingly favourable strip of habitat that organisms rarely leave; but this would require that the reduction in the per capita emigration rate be so large as to more than compensate for the increased population density. As described in Turelli (1994), both the prevalence of apparently stable hybrid zones in nature and Barton's (1979) mathematical heory lead us to expect that stable 'tension zones' between CI compatibility types can form along dispersal barriers. These tension zones can separate incompatibility types that show bidirectional incompatibility or form between predominantly infected and uninfected populations. They are likely to be important when attempts are made to use *Wolbachia* to introduce desirable genes into populations (Chapter 6).

2.6 Observed frequencies in natural populations

Despite many reports documenting the widespread occurrence of *Wolbachia* infections and CI in diverse insects, there are only two species (*D. simulans* and *D. melanogaster*) for which natural populations have been surveyed extensively. Only limited population data are available for *Culex pipiens*. Laven (1959) examined one strain per location in his geographic survey of CI in *C. pipiens* (Table 2.1). Subsequent investigations of Indian populations were the first to document intrapopulation variation in compatibility types (Subbarao *et al.* 1977*b*), but infection frequencies were not characterized. Barr (1980) described the distribution of incompatibility types in 69 egg rafts from southern California. Most (93 per cent) had the same incompatibility type, and there were no differences in CI type frequencies between populations. Magnin *et al.* (1987) reported reciprocal crosses among eight strains of *C. pipiens* collected within 100 km of Montpellier in southern France which indicated that all eight have different compatibility types. However, such data are insufficient to describe the fine-scale spatial distribution of types in nature and changes in their frequencies through time.

2.6.1 Frequencies in natural populations of *D. simulans*

2.6.1.1 *D. simulans* in California

As noted above, CI in *Drosophila* was first discovered in crosses between a predominantly infected population of *D. simulans* near Riverside in southern California and various uninfected populations in northern and central California (Hoffmann *et al.* 1986). Initially, the infection was apparently restricted to sites south of the Tehachapi transverse range that separates the Los Angeles basin from the Central Valley. However, infected flies became increasingly common in the Central Valley after 1988, and an extraordinarily rapid spread was observed within and among populations from 1989 until 1994. Because our experimental and theoretical analyses of this system have been very recently summarized (Turelli and Hoffmann 1995), only a brief overview will be presented.

This data set is notable for several reasons. It shows that CI-inducing infections can indeed spread within and among populations as predicted by the theoretical models discussed above. More striking is the quantitative agreement between the predictions of model (2.2) and the observed dynamics and apparent equilibria within populations. Repeated assays of females from nature have found that infected females are no less (or more) fecund than uninfected females. Hence, as expected from equation (2.2), the within-population dynamics seem to be adequately explained by considering only the levels of incompatibility between infected males and uninfected females, as measured by H, and the fidelity of maternal transmission, as measured by $1-\mu$. Our field estimates for these two parameterswith ranges $0.3 \leqslant H \leqslant 0.7$ and $0.03 \leqslant \mu \leqslant 0.05$—imply that once the infection reaches a local frequency of about 0.3 it should increase to near 0.8 within 7–43 generations. The average of our field estimates is $H = 0.55$ and $\mu = 0.04$; these imply that the increase from 0.3 to 0.8 should take approximately 14 generations. This prediction is consistent with the increases observed in four populations separated from each other by at least 100 km (Turelli and Hoffmann 1995).

A further prediction is that the infection frequency should increase only until the advantage from incompatibility is balanced by loss from imperfect maternal transmission. Using $0.3 \leqslant H \leqslant 0.7$ and $0.3 \leqslant \mu \leqslant 0.5$, equation (2.4) implies a stable equilibrium infection frequency between 0.85 and 0.99. With our average estimates $H = 0.55$ and $\mu = 0.04$, the predicted equilibrium is 0.94. Remarkably, 0.94 is the average frequency observed in five samples in Riverside from 1985 until 1993 and the average of five 1992 samples from the southern half of the Central Valley in which we had observed the increase of type R (Turelli and Hoffmann 1995). More recent data suggest that the frequency of infected flies may fluctuate as predicted by equation (2.4) with temporal changes in incompatibility and transmission levels.

In contrast to these cases in which both the dynamics and apparent equilibria show excellent quantitative agreement with model (2.2), we have also observed anomalous infection frequencies that are difficult to explain. The most extreme example comes from the Piru population, in the Tehachapi Range. It remained highly polymorphic in four samples from 1987, 1988 and 1989 (Hoffmann *et al.* 1990), before finally reaching near-fixation for type R in 1992. The observed polymorphic frequencies, 0.44–0.73, span a range over which type R should be rapidly increasing in frequency if the parameter values were near our average estimates. A possible explanation is that this represented a temporary tension zone between the highly infected populations of southern California and the uninfected populations of the Central Valley. Piru is in the middle of an extensive citrus-growing area with abundant resources for *Drosophila* year-round. Hence, the tension-zone hypothesis is plausible only if this area was characterized by extremely low emigration rates but high immigration rates.

The northward spread of type R was initially described by Turelli and Hoffmann (1991) as an example of a continuous 'Bartonian wave of advance'. Both the rate of northward movement and the width of the transition zone from highly infected to virtually uninfected populations suggested average dispersal distances in the order of 50 km/generation$^{1/2}$, far higher than plausible for dispersal by flight. We postulated human-mediated long-distance dispersal, associated with the transport of fruit throughout the state. In contrast to the cline of infection frequencies seen in 1990, an analysis of several northern California populations in 1992 showed that the frequency was essentially invariant over several hundred kilometres. This is different than the pattern expected from a travelling wave, and suggests that many northern populations may experience high levels of immigration each year from source populations to the south. Local extinction–recolonization can greatly facilitate the spread of infection across patchy habitats, because the resulting 'founder effects' allow individual populations to cross the unstable infection frequency threshold (cf. Lande 1979).

2.6.1.2 *D. simulans* outside California

Turelli and Hoffmann (1995) showed that *w*Ri is now prevalent in all North American populations surveyed, except for those in southern Florida. Although we were not able to follow the frequencies through time outside of California, our data from east-coast US populations strongly suggest that the infection frequency has increased fairly recently, probably over the past decade. Repeated samples (including unpublished 1995 data) from Columbia, South Carolina indicate that the infection frequency has remained for over a year near the apparent equilibrium value observed in California. We also found high frequencies of *w*Ri in samples from Costa Rica, Uruguay, Zimbabwe, Portugal, and Spain (the latter two unpublished). A 1994 sample from Israel indicated an intermediate infection frequency, suggesting that *w*Ri may currently be spreading through the Middle East (Turelli, unpublished). A highly polymorphic population was observed in France in 1990 (J. David, personal communication), and both *w*Ri infected and uninfected flies were captured in Italy in 1985 (Louis and Nigro 1989). Hence, we expect that many European populations should now be near the equilibrium described by equation (2.4). It would be useful to monitor the infection frequencies in some of these populations through time and estimate F, H, and μ to determine whether model (2.2) adequately describes the frequency dynamics of *w*Ri outside of California.

Much less information is available for the other incompatibility types in *D. simulans*. Incompatibility type H, associated with *w*Ha, was initially discovered in Hawaii (O'Neill and Karr 1990), but was subsequently found in other islands of the Pacific and Indian Ocean (Montchamp-Moreau *et al.* 1991), including the Seychelles (Rousset *et al.* 1992*a*; Merçot *et al.* 1995; Rousset and Solignac 1995). Rare uninfected flies have been found in Hawaii (Turelli and Hoffmann 1995), indicating that populations containing *w*Ha may reach incompatibility–transmission equilibria similar to that given by equation (2.4). Incompatibility type S, associated with double infection by *w*Ha and *w*No, has been found in the Seychelles and New Caledonia (Rousset *et al.* 1992*a*; Merçot *et al.* 1995). The appearance of some flies carrying only *w*Ha in a collection from the Seychelles suggests imperfect transmission of the double infection in nature (Merçot *et al.* 1995; Rousset and Solignac 1995).

Four separate areas—Australia, Florida, Ecuador, and Madagascar—are known to have *Wolbachia*-infected flies that neither cause incompatibility when mated to uninfected flies, nor protect females against incompatibility with type R, H, or S males (Rousset and Solignac 1995; Turelli and Hoffmann 1995; Hoffmann *et al.* 1996). It is difficult to understand the persistence of these flies in nature unless they produce some advantage, such as increased female fecundity (see equation (2.6) above). A similar dilemma arises when trying to understand infection frequencies in *D. melanogaster*.

2.6.2 Frequencies in natural populations of *D. melanogaster*

CI in *D. melanogaster* was first found in Australia (Hoffmann 1988), and CI-inducing *Wolbachia* infections have now been found in *D. melanogaster* from all continents (Solignac *et al.* 1994). As noted by Solignac *et al.* (1994), the average infection frequency of about 34 per cent from their temporally and spatially heterogeneous samples is difficult to reconcile with the simple equilibrium prediction (equation (2.4)). However, out of the many populations surveyed, Solignac *et al.* (1994) had sample sizes greater than 10 for only two. In one (Nasrallah, Tunisia), none of the 33 lines examined were infected; whereas in the other (Brazzaville, Congo), 14 out of 38 were infected. The latter produces an estimate of $p = 0.37$; but the corresponding 95 per cent confidence interval, (0.22, 0.54), does not exclude 0.5, the lowest equilibrium frequency consistent with equation (2.4). Hence, the Solignac *et al.* (1994) data do not preclude the possibility that the infection frequencies in *D. melanogaster* can be understood in terms of model (2.2), assuming that the species is divided into a patchwork of populations with and without the infection. To test this, larger samples and sequential samples from more natural populations are needed.

Hoffmann *et al.* (1994) surveyed 12 populations in eastern Australia with sample sizes ranging from 12 to 38 and found infection frequencies between 18 and 85 per cent. In three of the populations (all in the south), the infection frequency was significantly below 0.5, precluding an equilibrium described by equation (2.4). Hoffmann *et al.* (1994) found no evidence for reduced (or increased) fecundity of infected females. Values of $\mu = 0.02$–0.05 have been obtained in two experiments with wild-caught females (Hoffmann, unpublished). Assuming that $F = 1$ in equation (2.2), the constraints (2.3) for the existence of a stable polymorphic equilibrium described by (2.4) can be re-expressed as:

$$s_h > 4\mu(1-\mu). \qquad (2.8)$$

With $\mu = 0.02$–0.05, this requires $s_h > 0.08$–0.19 in nature. Hoffmann (1988) and Hoffmann *et al.* (1994) obtained laboratory estimates of s_h in the order of 0.0–0.3 using young *D. melanogaster* males from Australian populations. As in *D. simulans*, values of s_h in nature tend to be much lower than the laboratory estimates obtained with young males; the three estimates to date (Hoffmann, unpublished) indicate that $s_h = 0$. This suggests that these *Wolbachia* infections may confer fitness advantages that have not yet been identified. Long-term sampling is needed to test whether this infection is at an equilibrium in natural populations.

2.6.3 Cytoplasmic incompatibility and mitochondrial DNA

In *D. simulans*, different infections are closely associated with the three main mtDNA haplotypes found in this species (Table 2.4). Types R, W and (probably) A are associated with mtDNA type *si*II, whereas types H and S are associated with *si*I (Montchamp-Moreau *et al.* 1991). *Drosophila simulans* from Madagascar that carry an infection not causing incompatibility are associated with the third major mtDNA type found in *D. simulans* (Rousset and Solignac 1995).

Variation within the *si*II mtDNA type has also been linked to *Wolbachia*. The *w*Ri infection is associated with an *si*II variant (*si*IIB) that can be distinguished from another variant (*si*IIA) by a restriction fragment length polymorphism. The rapid sweep of type R northward through California produced a similarly dramatic change in these mtDNA haplotype frequencies. The *si*IIA type was initially prevalent in uninfected flies from California (Hale and Hoffmann 1990), but this changed as type R spread (Turelli *et al.* 1992). The frequency of the *si*IIB variant among uninfected flies became correlated with the frequency of the *w*Ri infection in a population (Turelli *et al.* 1992; Turelli and Hoffmann 1995); *si*IIB is rare in populations that are uninfected, whereas all uninfected flies carry this variant when the infection frequency is high. This is because when the infection is close to fixation, the only uninfected cytoplasm types present are those arising from the process of segregation, and these will be *si*IIB.

In *D. melanogaster*, there is no association between *Wolbachia* and mtDNA variation (Solignac *et al.* 1994). Because a spreading *Wolbachia* infection will

Table 2.4 Association between mitochondrial DNA and incompatibility types in *Drosophila simulans*

Incompatibility type	Mitochondrial DNA type
W	*si*II
R	*si*II
H	*si*I
S	*si*I
M	*si*III

also change the mtDNA of a population, this suggests that the infection in *D. melanogaster* is old and precedes divergence in mtDNA types. In contrast, the association between infection type and mtDNA variation in *D. simulans* suggests a more recent origin prior to divergence in mtDNA. Paternal transmission will break down any associations between an infection and mtDNA variation; thus, it is also possible that paternal transmission is more common in *D. melanogaster* than in *D. simulans*. Paternal transmission has not yet been investigated in *D. melanogaster*.

Incompatibility can cause rapid changes in mtDNA variation in laboratory populations. Nigro and Prout (1990) started two sets of *D. simulans* population cages carrying two mitochondrial variants (P and C), with one of the variants at a frequency of 20 per cent in one set and 80 per cent in the other. The C variant was infected by *w*Ri *Wolbachia* whereas P was uninfected. In all cages, there was a rapid increase in the frequency of the C variant as the infection became predominant under unidirectional incompatibility. In the mosquito, *Aedes albopictus*, changes in mtDNA frequencies in the laboratory have also been associated with a spreading infection (Kambhampati *et al.* 1992). In cages initiated with bidirectionally incompatible strains of mosquitoes, mtDNA haplotypes did not change consistently in frequency. However, in cages where there was unidirectional incompatibility, the mtDNA variant associated with the infected stock increased rapidly and replaced the other variant after two generations.

These findings have implications in two areas. First, they indicate that patterns of mtDNA variation in natural populations can be influenced by incompatibility. When populations lack mtDNA variation but contain high levels of nuclear variation, it is often assumed that a history of founder events has decreased mtDNA variation more than nuclear variation (Birky *et al.* 1989). However, this situation can also arise when cytoplasmic incompatibility results in the replacement of one mtDNA variant by another, without altering levels of nuclear variation (Hale and Hoffmann 1990; Turelli *et al.* 1992). The second implication concerns the detection of selection on mtDNA variants. When the frequencies of such variants change in populations faster than expected under random drift, this does not necessarily reflect selection on mtDNA. Instead, changes in mtDNA frequencies may reflect selection on other cytoplasmic factors such as *Wolbachia*. This point is discussed in Nigro and Prout (1990) and in Kambhampati *et al.* (1992) and has led to incompatibility being taken into account when testing for selection on mtDNA (e.g. Jenkins *et al.* 1996).

2.7 Evolutionary changes in *Wolbachia* and hosts

Prout (1994) presented the first rigorous theoretical analysis of the evolutionary fate of *Wolbachia* variants. He assumed perfect maternal transmission and complete compatibility between individuals infected with the alternative strains. Under these assumptions, he showed that selection favours the form conferring the highest fecundity on infected females, not the form producing the highest level of incompatibility between infected males and uninfected females. His analysis was generalized by Turelli (1994), who considered selection on both parasite and host genes that affect CI parameters. For mutually compatible *Wolbachia* strains, selection acts simply to maximize $F(1-\mu)$, the number of infected progeny produced by an infected female. Thus, contrary to intuition, selection among parasite variants does not act directly on the level of incompatibility with uninfected females. Variants that raise $F(1-\mu)$ are favoured even if they reduce incompatibility. Hence, there is evolutionary pressure for maternally inherited *Wolbachia* infections to become increasingly benign and possibly beneficial.

This simple picture is complicated if the alternative forms of *Wolbachia* that produce different values of $F(1-\mu)$ are not completely compatible with one another. Variants that raise $F(1-\mu)$ at the expense of being incompletely compatible with less benign variants will increase when rare only if they raise $F(1-\mu)$ sufficiently to overcome the progeny lost through incompatibility (see criterion 8 in Turelli 1994). The disadvantage for variants that do not satisfy this criterion depends on the composition of the population. In particular, variants that cannot increase when extremely rare may increase if they become sufficiently common through a founder event. Given that *Wolbachia* infections themselves are not expected to spread when rare if $F(1-\mu) < 1$—as is surely true for *w*Ri in *D. simulans*—the fact that *w*Ri has indeed spread means that we cannot necessarily restrict attention to modifiers that meet the criteria to increase when rare. This opens the door to more complex patterns of evolution that are difficult to predict without knowing the patterns of pleiotropic effects of *Wolbachia* modifiers. Nevertheless, the pressure to maintain compatibility with extant *Wolbachia* seems likely to maintain incompatibility between infected males and uninfected females.

The dynamics of modifier alleles in the host are difficult to predict, because such alleles will occur in both infected and uninfected individuals. No simple evolutionary criterion analogous to maximizing $F(1-\mu)$ emerges. Nevertheless, the relative fecundity of infected females compared to uninfected females, the efficiency of maternal transmission and the mutual compatibility of infected individuals all tend to increase under within-population selection on both host and parasite genes. Moreover, selection on host genes tends to favour increased compatibility between infected males and uninfected females. Thus, an important prediction is that levels of incompatibility may decrease through host evolution.

These predictions can be tested by transferring microbes between hosts and/or performing phylogenetic analyses of *Wolbachia* to distinguish ancestral from derived states, and some recent data seem to support them. For instance, the *Wolbachia* infection that causes very high levels of incompatibility in *D. simulans* causes much less incompatibility when transferred to *D. melanogaster* (Boyle *et al.* 1993), which also shows low levels of incompatibility produced by its 'native' *Wolbachia.* This suggests that the *melanogaster* genome is more resistant to *Wolbachia*-induced CI than the *simulans* genome, which would be consistent with the theory discussed above if the *D. melanogaster*–*Wolbachia* association is older than the *D. simulans*–*Wolbachia* association. The mtDNA analyses of Solignac *et al.* (1994) support this conjecture. In addition, the experimental transfer of *w*Ri from *D. simulans* to another *Drosophila* species (*D. serrata*), unlikely to have been exposed to *Wolbachia,* resulted in high levels of incompatibility and poor maternal transmission (Clancy and Hoffmann 1997). Such patterns are expected in a recently infected species where there has been no host selection.

In contrast to these observations, it is more difficult to reconcile predictions concerning *Wolbachia* evolution with the recent discovery of several infections that neither cause CI nor protect their carriers from it. It remains to be seen if these symbionts produce some benefits for their hosts to facilitate their spread or have a particularly high level of horizontal transmission. It is also possible that differences in CI between species have a non-adaptive explanation. For instance, the effects of *Wolbachia* on CI may depend on differences between species in sperm size or physiology rather than evolutionary changes in *Wolbachia* and its hosts. The sperm-length conjecture follows from the fact that *D. melanogaster* has longer sperm than *D. simulans* (1.91 mm versus 1.14 mm, Pitnick *et al.* 1995), suggesting that longer sperm may tend to produce lower levels of incompatibility. However, *D. recens* has much longer sperm than either (7.55 mm) but displays a level of incompatibility closer to *D. simulans* than to *D. melanogaster* (Werren and Jaenike 1995); whereas *D. ananassae* has intermediate sperm length (2.2–2.4 mm, Ashburner 1989, Table 8.4) and a relatively low level of incompatibility (Bourtzis *et al.* 1996). Nevertheless, it remains plausible that other physiological differences between species can affect incompatibility levels.

2.8 Conclusion

Where do we go next? The importance of single and multiple infections in causing incompatibility needs to be ascertained. Are the complex incompatibility patterns observed in *Culex* associated with multiple infections? Do multiple infections spread at the cost of single infections, or is this potential limited by poor maternal transmission? Testing for multiple infections in mosquitoes and monitoring infection frequencies in *Drosophila* populations carrying double and single infections should provide answers to these questions. The association between CI and multiple infections in other insects should also be examined, focusing on the many species known from molecular studies to carry such infections (Werren *et al.* 1995*a*, *b*).

The population studies on the *w*Ri infection of *D. simulans* indicate that infection frequencies in populations can be explained once field data on incompatibility and maternal transmission are obtained. However, additional information may be required to understand the population dynamics of other incompatibility systems. In *D. melanogaster*, infection frequencies in natural populations suggest that there is a beneficial effect associated with *Wolbachia*; this effect could be small and remains to be identified. In *Tribolium*, there is indirect evidence that *Wolbachia* affects sperm competition, although this effect was not detected in two *Drosophila* species. Additional experiments are needed to conclusively demonstrate *Wolbachia* effects on sperm competition. Studies on *Wolbachia* frequencies in natural populations of *Tribolium* and other insects may help to elucidate the importance of this factor and others such as larval curing. Population data from insect populations fragmented by reproductive barriers will be useful in testing whether 'tension zones' are common between infected and uninfected populations or between populations carrying different infections.

More research is needed to test the impact of *Wolbachia* infections on mtDNA variation. Patterns of mtDNA variation among populations are usually interpreted in terms of population histories and bottlenecks, without considering the importance of factors directly influencing the frequency of mtDNA variants. If *Wolbachia* infections generally influence such patterns, then insects with high levels of mtDNA variation among populations may often show strong intraspecific incompatibility.

Finally, we still dont know the molecular mechanisms by which *Wolbachia* causes incompatibility. While there is cytological information on the way the microorganism exerts effects on host chromosomes, it is not clear how sperm from infected males cause embryo death, how this sperm is modified by the presence of *Wolbachia*, or the mechanism by which the presence of *Wolbachia* in the young egg can 'rescue' *Wolbachia*-modified sperm back to normal function. The mechanism must involve highly specific *Wolbachia* products to account for the presence of bidirectional incompatibility and incompatibility among hosts carrying single and double infections.

Acknowledgements

We thank S. L. O'Neill and J. H. Werren for comments on previous drafts. This research was supported in part by grants from the Australian Research Council (A19531450) and the National Science Foundation (BSR 9119463, DEB 9527808).

3 Inherited microorganisms and sex determination of arthropod hosts

Thierry Rigaud

3.1 Introduction

The sexual differences between male and female animals are genetically determined by chromosomal sex factors, commonly carried by the sex chromosomes. Interestingly, the genetic basis of sex determination is highly variable, especially among arthropods. One of the classical modes of sex determination, heterogamety, occurs when one sex is heterozygous for the sex factors. When the male sex is heterozygous (male heterogamety), sex chromosomes are commonly designated X/Y and X/X in males and females respectively. In the case of female heterogamety, sex chromosomes are commonly designated Z/Z in males and W/Z in females. Heterogamety is probably the most common sex-determining mechanism known in diploid animals, and is almost always accompanied by anisogamy (unequal size of the gametes: males produce small gametes with little cytoplasm, while females produce large gametes containing a large amount of cytoplasm) (Bull 1983). A balanced sex ratio is maintained by Mendelian inheritance because one sex produces haploid gametes of different types (X and Y sperm in the case of male heterogamety) while the other sex produces only one type of gamete. Fertilization yields 1/2 XX and 1/2 XY zygotes, determining females and males respectively. In some animals, however, sex determination is complicated by the influence of environmental factors, e.g. temperature (in some reptiles, fishes and crustacea), photoperiod (in some crustacea), and crowding (in nematodes) (reviews in Ginsburger-Vogel and Charniaux-Cotton 1982; Bull 1983; Adams *et al.* 1987). These effects underscore the inherently labile nature of sex determination in many animal species.

Sex determination can also be affected by inherited intracellular microorganisms. While the advantage for a microorganism to live in a host can easily be understood in terms of securing a favourable habitat, the advantage of disrupting a mode of sex determination is not as clear. In the context of heterogametic sex determination coupled with anisogamy, a strict intracytoplasmic symbiote is at an advantage if it lives in a female host as opposed to a male host. This is because inherited microorganisms are predominantly transmitted vertically through the females egg cytoplasm, not the males sperm. The male represents a dead-end host for such a microorganism (Fig. 3.1a). The symbiont lineage will disappear with the death of its host, unless it is able to escape and be horizontally transmitted to another host. Therefore, any effect of the microorganisms metabolism which distorts the hosts sex ratio towards females will be selectively advantageous to the microorganism (Fig. 3.1b), provided of course that such a sex-ratio distortion does not unduly influence the ability of the host to reproduce.

An example of such a distortion is symbiont-mediated feminization of genetic males into functional females in some crustaceans. This increases the chance of the microbe being transmitted to the next generation by increasing the number of productive hosts. In fact this outcome is very similar to the effect of symbiont-induced parthenogenesis in *Trichogramma* (Chapter 4); the main difference is that feminizing microbes transform genotypic males into phenotypic females while microbes inducing parthenogenesis transform genotypic males into genotypic parthenogenetic females.

This chapter will focus on different symbionts which alter the sex determination of their host (called cytoplasmic sex factors, CSFs). We will see that their known distribution to date is limited to the Crustacea. The causes of this narrow distribution will be discussed, and the evolutionary consequences (known or suspected) for the host will be considered.

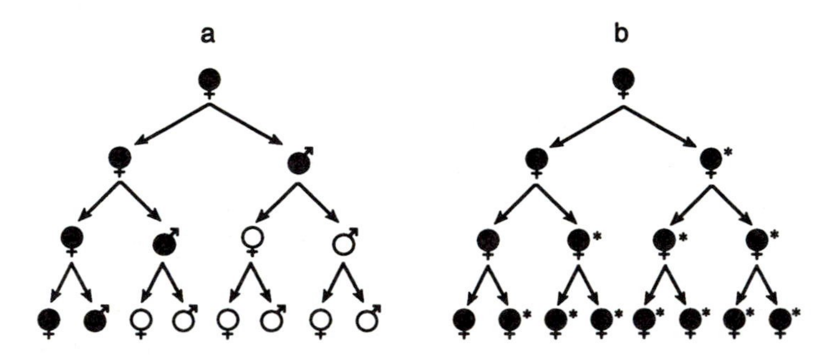

Figure 3.1 Vertical transmission of intracytoplasmic symbionts in the context of gonochory and anisogamy. (a) Transmission pattern of a symbiont without any effect on sex determination of the host. Males are a dead end since they do not transmit the symbiont. (b) Transmission pattern of a feminizing symbiont (* = genotypic males reversed into phenotypic females). The symbiont increases its fitness because it can be transmitted by all individuals. Filled symbols are infected and unfilled symbols are uninfected. At each generation, individuals are crossed with animals of the opposite sex and produce two offspring. To simplify, the sex ratio is fixed to 1 : 1 in uninfected individuals and in individuals infected with non-feminizing symbionts. The transmission rate of the symbionts is fixed at 1.

3.2 The incidence of cytoplasmic sex determination

3.2.1 Isopod crustaceans

Isopod crustaceans have been known for a long time to display aberrations in sex determination and sex ratio, both in the wild and in laboratory strains (Vandel 1941). The most intensively studied example of microbe-induced sex determination is found in the woodlouse, *Armadillidium vulgare* (Oniscidea) (Fig. 3.2). In this species, the genetic basis of sex determination is female heterogamety (♂ ZZ and ♀ WZ) (Juchault and Legrand 1972). Nevertheless, early genetic studies showed that some females regularly produced highly female-biased progenies without differential mortality between the sexes, and that this trait was maternally inherited (a phenomenon called thelygeny) (Vandel 1941; Howard 1942, 1958). A crucial finding for the understanding of these aberrations was made when Legrand and Juchault (1970) showed that biased sex ratios can be produced by WZ females after the inoculation of tissues from thelygenic females. This suggested that an infectious agent could be responsible for sex-ratio bias. The agent was discovered a few years later (Martin *et al.* 1973), and appeared to be a small endocytoplasmic bacterium living in the cells of all tissues in some females, but never found in males.

The phylogenetic status of these bacteria was first suspected by their morphology and life-cycle characteristics (Rigaud *et al.* 1991*a)*, and then confirmed by molecular sequencing techniques (Rousset *et al.* 1992*b*). They belong to the genus *Wolbachia*, a monophyletic group belonging to the α subdivision of the Proteobacteria, and are therefore close to symbionts causing cytoplasmic incompatibility or parthenogenesis in several insect species (O'Neill *et al.* 1992;

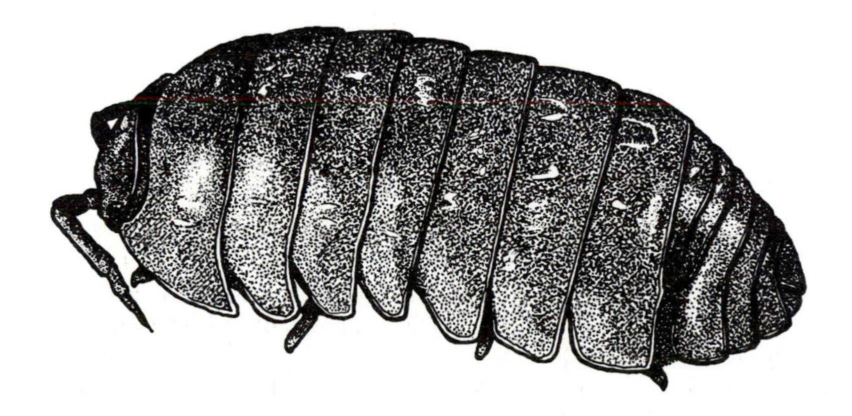

Figure 3.2 *Armadillidium vulgare* female (1 cm actual length) (drawing: T. Rigaud).

Rousset *et al.* 1992*b*; Werren *et al.* 1995*a*). *Wolbachia* from *A. vulgare* are closely related to those of *Culex pipiens*, *Nasonia vitripennis* and *Trichogramma* sp., and belong to the B subdivision of the *Wolbachia* genus as defined by Werren *et al.* (1995*a*). In the woodlouse, *Wolbachia* appear to be quite thermosensitive; temperatures up to 30 °C lead to destruction of the bacteria in the hosts tissues (Juchault *et al.* 1980*b*). As a consequence, very young *Wolbachia*-infected females of *A. vulgare* reared at 30 °C progressively acquire a male phenotype (Juchault *et al.* 1980*b*). When older infected females are reared at 30 °C, they are unable to revert into males, presumably because their sexualized tissues are terminally differentiated. However, when these older females are crossed, they produce highly male-biased broods, instead of the highly female-biased broods commonly observed when individuals are reared at 20 °C (Rigaud *et al.* 1991*b*).

These results imply that infected females possess an atypical female genotype, since 'normal' WZ females infected with *Wolbachia* would be expected to produce 1♂ : 1♀ broods at 30 °C. Furthermore, there would be no expectation for a young WZ infected-female to develop into a male after exposure to 30 °C (in fact, temperature has no effect on the WZ uninfected female phenotype [Juchault *et al.* 1980*b*]). These results can be easily interpreted by assuming that all infected females are of a male genotype (ZZ), and that they possess a female phenotype only because of the symbionts presence, resulting in a feminizing effect due to the presence of *Wolbachia*. This feminizing effect of the inherited microorganism can explain all phenomena so far observed in infected strains of *A. vulgare* (Box 3.1). Finally, different antibiotics also cure *Wolbachia* from host tissues and the concomitant effect is suppression of the feminizing activity (Rigaud *et al.* 1991*a*).

In populations where *Wolbachia* occur, all infected females have been found to be ZZ (genotypic males) sexually reversed by the bacteria (Juchault *et al.* 1980*a*, 1993). In infected strains, the sex determination of the isopods is therefore under the control of the symbiont: individuals inheriting *Wolbachia* develop into females, while males are uninfected. As such, the bacteria of *A. vulgare* are an example of a cytoplasmic sex factor (CSF). Symbiont transmission is about 90% efficient (most infected females produce only daughters) and the sex ratio in broods of infected females is consequently largely female-biased (Juchault *et al.* 1980*a*, 1993). The microbe-induced feminization can nevertheless be incomplete, in a few cases leading to intersex phenotypes (Legrand and Juchault 1969; Legrand *et al.* 1987). These phenotypes vary from functional females with tiny male sexual secondary characters to sterile individuals possessing sexual characteristics of both sexes and hermaphrodite non-functional gonads. This gradation is suspected to be related to a decreasing bacterial density in individuals (Rigaud *et al.* 1991*a*), or to a delay in bacterial activity (Rigaud and Juchault 1993; see section 3.3.1).

Box 3.1 Crosses involving *A. vulgare* females infected by a cytoplasmic sex factor (CSF) at two temperatures

(a) ♂ZZ × ♀WZ + CSF —20 °C→ **♀WZ + CSF; , ZZ + CSF;** (♀WZ); (♂ZZ)

(b) ♂ZZ × ♀ZZ + CSF —20 °C→ **♀ZZ + CSF;** (♂ZZ)

(c) ♂ZZ × ♀ZZ + CSF —30 °C→ **♂ZZ** (♀ZZ + CSF)

(a) Cross between a male and an infected genetic female, at a temperature alowing CSF survival
(b) Cross between a male and an infected genetic male reversed into functional female, at a temperature allowing CSF survival.
(c) Cross between a male and an infected genetic male reversed into functional female, at a temperature lethal for CSF.

Phenotypes in bold are in high proportion (usually >90 per cent), while those within parentheses are in a minority. The relative proportions of each phenotype depend on the CSF transmission pattern. Females of *A. vulgare* are normally heterogametic.

The presence of *Wolbachia* in the Isopoda is not restricted to *A. vulgare*. At least 17 other species of isopods belonging to different families are infected by *Wolbachia* (Bouchon and Rigaud, unpublished results), but the effect of these symbionts has not been determined in all cases. A feminizing effect is nevertheless known in *Armadillidium nasatum* (Juchault and Legrand 1979) and is strongly suspected in other infected species: *A. album*, *Ligia oceanica* (Juchault *et al.* 1974), *Chaetophiloscia elongata* (Juchault *et al.* 1994), *Porcellionides pruinosus* (Juchault *et al.* 1994; Rigaud *et al.* 1997), and in the estuarine isopod *Sphaeroma rugicauda* (Heath and Ratford 1990; Martin *et al.* 1994). One case of *Wolbachia* inducing cytoplasmic incompatibility is also known in *Porcellio dilatatus* (reviewed by Rigaud and Rousset 1996).

3.2.2 Amphipod crustaceans

The phenomenon of thelygeny and/or intersexuality is also known in the amphipods *Gammarus duebeni* (Bulnheim and Vavra 1968), *Orchestia gammarellus* (Ginsburger-Vogel 1975), *O. mediterranea* and *O. aestuarensis* (Ginsburger-Vogel 1991). These cases are similar to isopods, in that some females are thelygenous; this trait is associated with vertically transmitted microorganisms and feminization occurs in juveniles. However, a number of differences between isopods and amphipods exist. First, the causative agents of feminization are different. While isopods are feminized solely by *Wolbachia* bacteria, feminization is associated with microsporidia in *G. duebeni* (Bulnheimand Vavra 1968; Smith and Dunn 1991), and with a paramixydia (a protist closely related to microsporidia) in *O. gammarellus* (Ginsburger-Vogel and Desportes 1979). Secondly, *Wolbachia* and paramixydia are scattered throughout the hosts tissues (Martin *et al.* 1973; Ginsburger-Vogel and Desportes 1979), whereas the microsporidia are present only in the female germline. This difference may be due to a strategy of differential segregation in microsporidia, which seem unable to divide rapidly enough to follow the rapid divisions of the host's cell cycle, and therefore preferentially infect the target (transmitting) tissue (Dunn *et al.* 1995; Hatcher *et al.* 1997). The symbionts may be capable of identifying putative germline tissues and infecting them (Hatcher *et al.* 1997), but a differential replication rate of the symbionts in somatic v. germinative tissues may also be responsible for the selective repartition of microsporidia. Finally, the native sex-determining system is different in amphipods when compared to *A. vulgare*. Sex determination in uninfected *O. gammarellus* is male heterogamety (Ginsburger-Vogel and Magniette-Mergault 1981*a*, *b*). In *G. duebeni*, sex determination is controlled by a polyfactorial system (Bulnheim 1978) and also by the environment (Adams *et al.* 1987). The cue for this environmental sex determination is photoperiod (Naylor *et al.* 1988). However, in all cases, the genetic or environmental cues for sex are overridden by the feminizing effect of the symbionts and infected animals are feminized.

The feminizing agents of amphipods are also susceptible to environmental effects. Both microsporidia and paramixydia are affected by temperature variations (Bulnheim 1978; Ginsburger-Vogel and Magniette-Mergault 1981*a*, *b*), and the symbionts of *G. duebeni* are also affected by high water salinity (Bulnheim 1978).

3.2.3 Other crustaceans

There are numerous reports of female-biased sex ratios or intersexuality in crustaceans. These traits sometimes appeared to be controlled by simple Mendelian systems (Sassaman and Weeks 1993), polygenic systems (Battaglia 1963), or the environment (Ginsburger-Vogel and Charniaux-Cotton 1982; Adams *et al.* 1987). In addition to the well-studied examples of CSF described above, a cytoplasmic influence on sex determination has been demonstrated in the copepod *Tigriopus japonicus* (Igarashi 1964*a*, *b*). A cytoplasmic substance or symbiont (of unknown nature) may control the duration of the incubation of the eggs, which in turn influences the sex of the larvae. Male feminization by protozoans has also been reported in the decapod crab, *Inachus dorsettensis* (Smith 1905), and intersex phenotypes in the crab, *Leptomithrax longipes*, could be due to a bacterium present in the haemolymph and tissues (Roper 1979). In these last two cases, the transmission patterns of the microorganisms are unknown, and further work is necessary to confirm the presence of a CSF.

3.2.4 What explains the narrow niche of cytoplasmic sex determination?

Based on our present state of knowledge, cytoplasmic disruption of sex determination seems to be restricted to crustaceans. CSFs have also been suspected in a few insects and aphids (reviewed by L. D. Hurst 1993), but more information is needed to confirm this hypothesis. This raises the question of why feminization-induction by symbionts is not more widespread in taxa other than Crustacea. The first point to consider is the observation that there is more than one causative agent for feminization. This situation contrasts with that of cytoplasmic incompatibility, where the only known causative agent is *Wolbachia* bacteria, whatever the host's taxonomic position (Chapter 2). The fundamental basis of feminization must therefore be largely a property of the host rather than the microorganisms.

The occurrence of symbiont-induced feminization has probably been facilitated in crustaceans by their labile system of sex determination and differentiation. It appears that both male and female crustaceans possess all the genes necessary for differentiation of either sex, i.e. males are very easily converted into females and females into males by simple experimental manipulations (Charniaux-Cotton and Payen 1985; Legrand *et al.* 1987). Juchault and Mocquard (1993) proposed that males and females of *Armadillidium vulgare* share common genotypes, except that the female chromosome (W chromosome) is a male chromosome (Z chromosome) carrying an additional 'female gene' which suppresses the expression of 'male genes'. In fact, the cue for sex determination and differentiation in crustaceans seems to be the male gene(s) carried by the male chromosome(s). This gene(s) controls the development of the androgenic gland, the organ responsible for male hormone synthesis. In amphipods and isopods, this circulating hormone induces male sex differentiation after the third and fourth moult, respectively (Juchault 1966; Charniaux-Cotton and Payen 1985). This androgenic hormone was clearly identified in *A. vulgare* as a protein of 17–18 kDa (Martin *et al.* 1990). In the absence of the hormone, a female phenotype and a female physiology are autodifferentiated (Charniaux-Cotton 1959). From the compilation of empirical data, Legrand *et al.* (1987) proposed that the female gene in the case of female heterogamety inhibits the expression of the male gene(s), resulting in female differentiation.

In order for a symbiont to induce feminization, the differentiation of the androgenic gland must be blocked. In a way, 'feminization' is an inappropriate term since the effect of the inherited symbiont is more an inhibition of the development of the male phenotype rather than an induction of the female phenotype. The ease of feminization is illustrated by the parasitic castration induced by crustacean parasites in a number of decapods (Giard 1886; Veillet and Graf 1959). This castration results in androgenic gland degeneration, and induces a feminization of the tissues of infected individuals. Degeneration is related to the parasites presence in the central nervous system, and should therefore not be considered the result of direct contact between the androgenous gland and the parasite (Rubiliani *et al.* 1980).

The precise molecular mechanism by which feminizing microorganisms influence host sex determination is unknown. However, as is the case in chromosomic females, the androgenic gland never differentiates in ZZ females having inherited *Wolbachia* vertically (Legrand *et al.* 1987). This suggests that the symbionts are able to down regulate 'male gene(s)' activity, producing the same ultimate effect as that of the 'female gene(s)' carried by the female chromosome. While the molecular mode of action of the chromosomal and cytoplasmic sex factors may be acting at different points in a gene expression cascade, the ultimate phenotypic outcome is the same. From a purely genetic perspective, the major difference between these sex factors is their inheritance pattern (Smith and Dunn 1991). In addition, with the avoidance of androgenic gland differentiation, the *Wolbachia* possess a second feminizing property—a 'physiological effect'—perceivable when they are inoculated into adult males (i.e. individuals where androgenic glands are differentiated and functional). In this case, males acquire an intersex phenotype by differentiating female genital appertures and other female sexual characters (they are therefore 'feminized'). The androgenic gland of these males is still functional but undergoes progressive hypertrophy. The androgenic hormone is still synthesized and released, but is inefficient. As the hormone of these individuals remains active when inoculated into other individuals (Juchault and Legrand 1985), the effect of *Wolbachia* is not an inactivation of the hormone. The effect of *Wolbachia* may instead involve competition of a bacterial product and the hormone for the receptors to this hormone on the target tissues (Juchault and Legrand 1985). The two feminizing effects of *Wolbachia* must be considered separately because the first one avoids the differentiation of the androgenic gland and the other only occurs when this gland has been differentiated.

These effects of perturbing sex differentiation have not been found in insect—symbiont associations. In *Drosophila*, for example, no single genetic change is known that converts an XY male into a functional female, probably because of several factors being involved. One can invoke the lack of circulating sexual hormone, the 'cell-to-cell' sex differentiation (the phenomenon of clonal autonomy: the sex of one patch of tissue is not influenced by the sex in other parts of the fly [see Ferveur *et al.* 1995 for a case of partial feminization]), and the strong constraints of dosage compensation (Bull 1983; Wilkins 1993).

3.3 Evolutionary consequences of the presence of feminizing symbionts: an illustration of intragenomic conflicts

3.3.1 Evolution of sex-determining mechanisms: the case study of *Armadillidium vulgare*

The presence of a CSF will have consequences for the reproductive biology of the host. The most direct effect of a feminizing factor, as already pointed out, is the change in the inheritance pattern of the female sex determinant (from nucleus to cytoplasm). In organisms with basic heterogametic sex determination, such as *A. vulgare*, this occurs by the elimination of the chromosome bearing the female sex-determining gene within populations with the CSF by a simple selective process. Half the offspring of the 'genetic' (or chromosomal) females are daughters, while *Wolbachia*-infected females produce almost all daughters. Assuming an equal fecundity of both infected and uninfected females, the infection is therefore presumed to increase in the population from generation to generation, and genetic females can be entirely replaced by feminized males in a few generations (Bull 1983; Taylor 1990; Fig. 3.3). In species with female heterogamety, the W chromosome is rapidly eliminated, thus the nuclear female determinant is replaced by the cytoplasmic one (i.e. the *Wolbachia*). In *A. vulgare*, observations in the field are in accordance with this prediction, since all *Wolbachia*-infected females captured possess a male genotype (ZZ), and since several field populations do not harbour WZ females (Juchault *et al.* 1980*a*, 1992, 1993) (see Fig. 3.5).

However, the occurrence of CSFs does not mean that host genes are uninvolved in sex determination. As explained by Werren (1987) and Hurst (1992), cytoplasmic sex-ratio distorter genes are in conflict with the autosomal genes of the host because of their different inheritance patterns (nuclear genes tend to be inherited by both parents while cytoplasmic genes are only maternally inherited). Furthermore, as pointed out by Juchault *et al.* (1993), this conflict is also perceptible at the population level: although the microorganism will tend to produce 100 per cent females in the infected lineages, selection acting on chromosomal genes will tend to maintain the male sex which is necessary for reproduction. This conflict can be solved by two distinct evolutionary patterns. First, a coevolutionary feedback on sex ratio can be selected. Because of Fisherian selection favouring an overall 1 : 1 sex ratio, in the presence of a feminizing parasite a male-biased sex ratio may be selected in uninfected lineages (Werren 1987; Hatcher and Dunn 1995). The extent of this feedback will depend on the degree of sex-ratio coevolution (Werren 1987) and on the amount of feminizing activity of the parasite (Hatcher and Dunn 1995), but always requires the incomplete transmission of the feminizing element. The underlying assumption for this solution of genomic

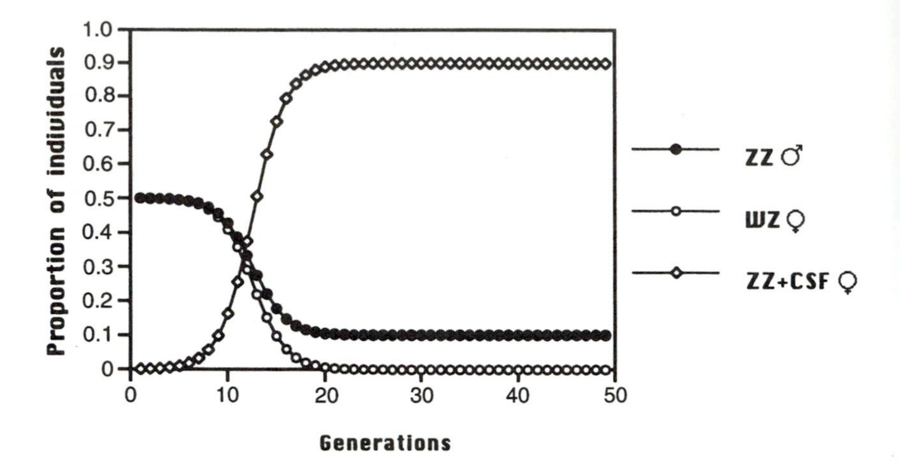

Figure 3.3 Plot of genotype frequencies versus generations in a hypothetical population infected by a vertically transmitted cytoplasmic sex factor (CSF). The infected ZZ + CSF females were introduced in the uninfected population at a rate of 0.01. The transmission rate of the feminizing factor was 0.9. After 20 generations, the population consisted only of ZZ individuals and the sex ratio (proportion of females) equals the symbiont transmission rate. (Equations for this plot came from Taylor 1990.)

conflict is that the sex-determining mechanism allows for heritable variation in sex ratio. In the simplest case of heterogamety, sex-ratio selection is difficult to obtain because sex-determining genes are limited in number and are fixed on a given chromosomal pair. In this case, however, the conflict can be solved by a different mechanism, i.e. the selection of autosomal genes of the host resisting the sex-ratio distorter. Such genes (probably a polygenic system) were observed in *A. vulgare* by selecting an infected strain where females produced male-biased broods (Rigaud and Juchault 1992). Their main effect is to limit the bacterial transmission to offspring, and they do not override the feminizing activity of the symbiont. Nevertheless, by modifying the *Wolbachia* transmission pattern, such genes intervene indirectly in sex determination.

Another way to solve the intragenomic conflict may occur in a different woodlouse, *Porcellionides pruinosus*. In this species *Wolbachia* also mediate a feminizing activity (Juchault *et al.* 1994); all individuals from all populations tested were infected, including males. This contrasts with the *Armadillidium* situation where males are never infected. In *P. pruinosus*, it has been suggested that some male-restorer genes exist. These genes would not affect the bacterial transmission as in *A. vulgare*, but would prevent the expression of *Wolbachia* and restore a male phenotype even in the presence of the feminizing factor. An alternative explanation could be that these *Wolbachia* have a slower growth rate than the strain in *A. vulgare* or could be weak expressors.

Another route for the evolution of sex determination has been observed in *A. vulgare*. The first trace of this evolutionary pattern was the finding of numerous female-biased strains lacking *Wolbachia* symbionts (Juchault and Legrand 1976*a*). These lineages possess traits similar to those of *Wolbachia*-infected strains, but also several traits distinguishing them (Box 3.2). The feminizing agent responsible for sex-ratio deviations in these strains was labelled f. Although the precise nature of the f factor still remains unsolved, its transmission or expression pattern strongly resembles those of self-replicating elements such as transposable elements responsible for hybrid dysgenesis (Bregliano *et al.* 1980; Anxolabéhère and Périquet 1983), and that of the σ virus in *Drosophila* (L'Héritier 1962). This comparison allowed Legrand and Juchault (1984) to propose that f could be a nuclear mobile element carrying feminizing information. The best argument for this hypothesis is that the f factor is not inherited following a Mendelian pattern (a nuclear gene with incomplete penetrance could also have been proposed as a candidate for f, but even if incomplete penetrance can explain different levels of expression in the feminizing effect, such a gene would always have a Mendelian inheritance). What is the link between this f factor and symbiont-induced sex-determination evolution? During five generations of studying the descent of one WZ female inoculated with *Wolbachia*, Legrand and Juchault (1984) observed the spontaneous appearance of a f lineage. This occurred after *Wolbachia* failed to be transmitted to a particular female. The f factor could not have come from males of foreign lineages since the lines were maintained by inbreeding. There was thus a direct link between the prior infection by *Wolbachia* and the appearance of f. Legrand and Juchault (1984) proposed that f could be either a phage associated with the *Wolbachia* or a part of the bacterial DNA incorporated into the nuclear genome of the isopod. Despite intensive research at the ultra-microscopic level, no viral particles were detected in the f-carrying females, and the hypothesis of f as a transposon-like factor carrying bacterial feminizing information remains likely.

Nevertheless, two alternative hypothesis can be proposed. One can first consider that the feminizing function of f need not be of bacterial origin. The f element could be the natural female-determining locus captured by a bacterial transposable element. The number of copies of this element or differences in its location (in heterosomes or in autosomes) could explain the variability in the expression of feminization. The other possibility could be that the transposable element has no feminizing property at all, but disrupts the male differentiation by inserting into the male gene(s). This inactivation of the male gene would be sufficient to allow the expression of the female phenotype. Under these hypotheses, the disappearance of bacteria (non-transmission) is necessary to observe the effects of the f factor because *Wolbachia* would otherwise mask them. Clearly, further molecular genetic studies are needed to choose among the speculative hypotheses described above about the nature of the f element. Nevertheless, whatever the underlying mechanism, the symbiotic

Box 3.2 **Characteristics of lineages with the f factor**

Similarities with *Wolbachia*-infected strains:

- The inheritance of the female-biased sex ratio is mainly maternal (Juchault and Legrand 1976*a*).
- High temperatures restore male-biased sex ratios (Legrand and Juchault 1984).
- Females are genetic males reversed by a feminizing factor (Legrand and Juchault 1984).

Differences from *Wolbachia*-infected strains:

- No microorganisms present in f lineages.
- The female-biased trait can occasionally be transmitted by males in f lineages, a phenomenon never observed in the *Wolbachia*-infected strains despite the numerous breeding experiments in the lab (Legrand and Juchault 1984; Juchault *et al.* 1992, 1993). However, the inheritance of male transmission is very unstable, showing a non-Medelian pattern (Legrand and Juchault 1984).
- The inheritance of the sex ratio is very unstable. By following iso-female lines individually, one can observe wide sex ratio variation according to generations (varying from all-female broods to almost all-male broods), which is never observed in *Wolbachia*-infected lineages. In addition, a decrease in the female proportion can be observed in the successive broods of a single mother (owing to their capacity to store sperm, female woodlice can produce several broods without remating). In this case, sex ratio can evolve from 100 per cent daughters to 100 per cent sons over only three broods (Legrand and Juchault 1984), as if there was a progressive exhaustion in the feminizing factor transmission or expression during the mother's ageing.

association between *Wolbachia* and the isopods seems to be the origin of this new sex-determining mechanism (Fig. 3.4).

Because of its non-Mendelian inheritance, the presence of the f factor also leads to intragenomic conflict, in a way very close to that induced by the presence of CSF (see above). This conflict is visible at the population level: an autosomal masculinizing gene can restore the male sex in the presence of f, but is inefficient against the feminizing effect of *Wolbachia* (Rigaud and Juchault 1993; Box 3.3). The high rates of intersexes in broods of *Wolbachia*-infected females (Rigaud and Juchault 1993) nevertheless indicate a slight effect of this *M* gene on the bacterial feminizing effect. These intersexes begin their sexual development by a short period of male differentiation, rapidly followed by a period of female differentiation (Juchault 1966). As a consequence, they are

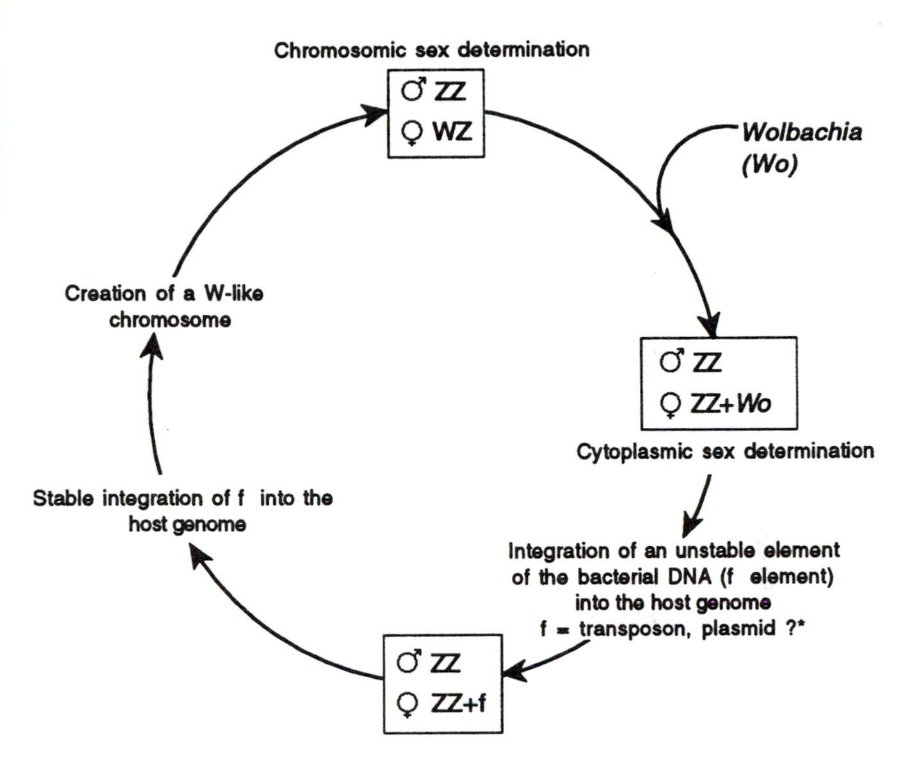

Figure 3.4 Schematic view of the evolution of sex-determining mechanisms in *Armadillidium vulgare*, modified from Juchault and Mocquard (1993); *, see text for alternative hypotheses on the nature of the f element.

functional females (they produced viable offspring as mothers), but possess tiny non-functional external male appendixes. The prior male differentiation indicates that the *M* gene was first able to override the feminizing effect of *Wolbachia*. But, in the very early stage of this male differentiation, the second feminizing effect of *Wolbachia* (the physiological effect, see section 3.2.4) stops the male differentiation by inactivating the androgenic hormone. This effect is similar to that observed after the inoculation of *Wolbachia* in adult males (Rigaud and Juchault 1993). One can speculate that the secondary 'physiological' effect of *Wolbachia* could have been selected in response to a resistance gene of the *M* type. In accordance with Fisher's predictions, this gene is not selected in populations where sex determination is chromosomal (Juchault and Legrand 1981*b*), but it can be present at high frequencies in populations influenced by feminizing factors. Interestingly, its frequency increases with the frequency of the f factor (Juchault *et al.* 1992). This evolutionary change may not be an isolated case: in at least one other woodlouse species (*Porcellio scaber*), non-Mendelian sex-ratio deviations are known in strains free of *Wolbachia* infection (Juchault and Rigaud, unpublished results).

Sex-determining mechanisms in *A. vulgare* are capable of evolving further. The f factor, under unknown conditions, is able to acquire a stable Mendelian inheritance pattern in a limited number of lines (Juchault and Mocquard 1993). Using crossing experiments, it was shown that this stable fixation of f occurs on a male chromosome (Z chromosome), and creates a W-like chromosome. The fixed female gene can not be anything but the f factor, because the lineages used for this experiment were maintained in the lab for many years, and were known to harbour only this feminizing element (Juchault and Mocquard 1993). This observation confirms the small differences between male and female sex chromosomes: the only difference between them would be the presence or absence of the female determinant, the rest of the chromosome being identical (Juchault and Mocquard 1993; see also section 3.3.3).

This observation also suggests a circular view of the evolution of sex-determining mechanisms in *A. vulgare* (Fig. 3.4), and one can wonder about the original starting point of this evolutionary pattern. It is commonly assumed that native sex determination is nuclear in crustacean species and influenced by feminizing factors (Legrand *et al.* 1987). However, cytoplasmic effects on the evolution of sex-determining mechanisms and/or reproductive systems have been considered in other taxa (e.g. plants), where nucleo-cytoplasmic

Box 3.3 **Effects of a putative 'masculinizing' autosomal gene on feminizing sex-ratio distorters**

(a) ♂ZZ *Mm* × ♀WZ *mm* → 1/4 ZZ *mm* (♂); 1/4 ZZ *Mm* (♂); 1/4 WZ *Mm* (♂); 1/4 WZ *mm* (♀)

(b) ♂ZZ *Mm* × ♀ZZ *mm*+f → 1/2 ZZ *mm*+f (♀); 1/2 ZZ *Mm*+f (♂)

(c) ♂ZZ *Mm* × ♀WZ *mm*+B → 1/2 ZZ *mm*+B (♀); 1/2 ZZ *Mm*+B (i or ♀)

B, feminizing bacteria (*Wolbachia*); f, feminizing f factor; i, intersexes; *M*, putative dominant masculinizing autosomal gene. The transmission rate of feminizing factors is assumed to be 100 per cent. This transmission rate can nevertheless be variable, inducing variations in the proportion of male and female phenotypes.

A masculinizing gene was first suggested when wild-caught males were crossed with genetic females (cross a) (Juchault and Legrand 1976*b*). In the present of both f and *M*, individuals develop into males (cross b), as is the case when *M* and the W chromosome are in competition. The *M* gene is, however, unable to restore the male sex in the presence of the feminizing *Wolbachia* (cross c).

interactions may have favoured the evolution from hermaphroditism to dioecy (existence of separate sexes) (Maurice *et al.* 1994). It is therefore interesting that the evolution of sex determination in *A. vulgare* could be a relict of a more ancient evolution in crustacea, i.e. the appearance of separate sexes (thus the appearance of sexual chromosomes) from a hermaphroditic ancestor, via cytoplasmic sex determination (Juchault and Mocquard 1993). Even though speculative, this hypothesis may have merit, since most of the primitive crustaceans are hermaphroditic (Hessler *et al.* 1987), and we know (section 3.2.4) that it is very easy to obtain male or female phenotypes in crustaceans with relatively few genetic changes (Charniaux-Cotton and Payen 1985).

3.3.2 The population distribution and impact of feminizing sex-ratio distorters

Most studies of feminizing factors have been undertaken on lines maintained in the lab. The main reason for this is that long rearing studies are necessary to observe and analyse the phenomenon (maternal inheritance, evidence of a link between the symbionts and the sex-ratio bias), in part because most crustaceans have long life cycles. However, some studies have focused on the prevalence and influence of feminizing factors in field populations.

The most complete data set comes from the woodlouse *Armadillidium vulgare* (Juchault and Legrand 1981*a*,*b*; Juchault *et al.* 1993). The results of field sampling over 15 years are summarized in Fig. 3.5. It appears that the most common sex factor found in populations is the f factor. The prevalence of f nevertheless varies between populations, and ranges from 3 to 100 per cent when present. The least common female sex factor is *Wolbachia*: most populations lack symbionts, and, when present, they are often at a low frequency (from 6 per cent to the exceptional case of 70 per cent). In contrast, field collections of *Gammarus duebeni* showed that all populations studied were infected by feminizing microorganisms (microsporidia). The levels of infection varied between populations, from 5 to 30 per cent (Bulnheim and Vavra 1968; Dunn *et al.* 1993*a*; Dunn, personal communication). In *Orchestia gammarellus*, Ginsburger-Vogel and Desportes (1979) found that, on average, estuarine populations are more likely to be infected than seashore populations. But even in infected populations, the frequency of feminizing paramixydia does not reach 100 per cent.

Three major points are raised by these distributions:

(1) Why do CSFs not occur in a majority of individuals in the populations they infect, and why are they not more widespread?

(2) How can we explain the maintenance of the female sexual chromosome (or other basic sex-determining systems) in so many populations, while it should be eliminated by the feminizing factors (see section 3.3.1)?

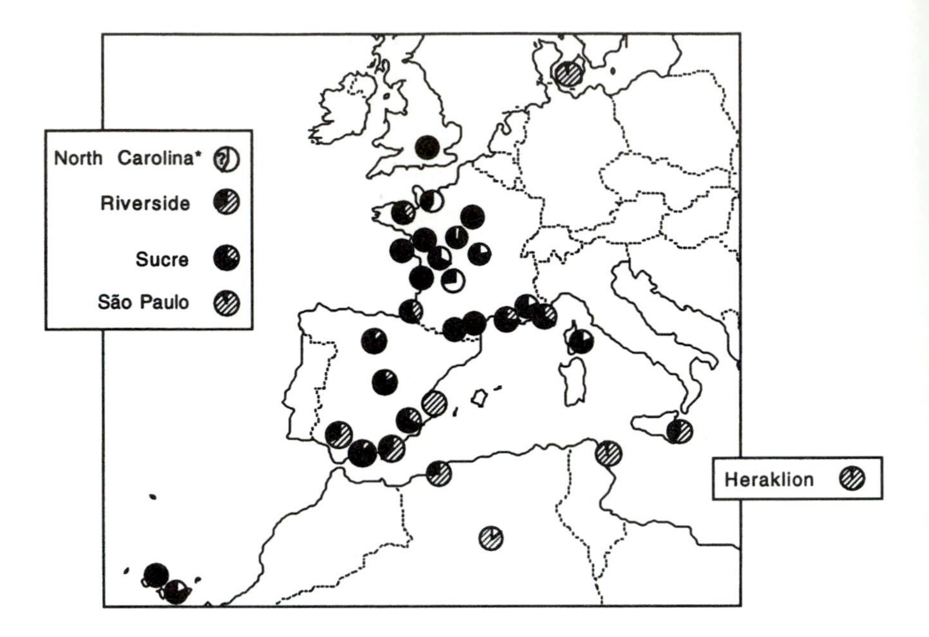

Figure 3.5 Distribution and frequency of sex factors in wild populations of *Armadillidium vulgare*. ●, Females harbouring the f factor; ○, females harbouring the feminizing *Wolbachia*; ◍, genetic females (sex determined only by the W chromosome); *, incomplete information, these half females (denoted '?') did not harbour *Wolbachia*, but the authors were unable to determine whether they had the f factor. (Data from Ganter and Hanton 1984; Juchault *et al.* 1993; and unpublished results.)

(3) Why is the feminizing *Wolbachia* rarer than the f factor in *A. vulgare* populations?

It seems likely that answers to these questions have to involve phenomena specific to each system and one wonders if a general explanation exists. The relatively low symbiont prevalence could be due to a fitness cost to the host. In *Gammarus duebeni*, such a negative effect has not been observed on the major life-history traits by comparing the performances of infected v. uninfected females (growth, mortality, reproductive success) (Dunn *et al.* 1993*a*). In *A. vulgare*, Juchault and Mocquard (1989) showed that when *Wolbachia* are inoculated into uninfected females, the result is a decrease in their growth rate in the months following the inoculation, with the result that females reach the reproductive period with a lower weight than their uninfected counterparts, and therefore produce fewer offspring. This decrease in fitness can be large enough to reduce the probability of the spread of an infection of this kind in a population (Rigaud *et al.* 1992). However, this cost seems to disappear in subsequent generations of the infected strain (Rigaud, unpublished results). Infected or uninfected females captured in the wild produce equivalent numbers of offspring at equivalent weight. Juchault and Mocquard

(1989) nevertheless observed that the older females in the wild are uninfected, and are therefore more likely to produce more offspring than infected females.

Intersexes occur in all the species with CSF (Ginsburger-Vogel and Desportes 1979; Juchault and Legrand 1981*b*; Dunn *et al.* 1994). While intersexuality is not a specific consequence of CSF, because it can also be induced by environmental sex determination (Dunn *et al.* 1993*b*), the presence of such sterile individuals at rates from 5 to 30 per cent can be a serious load for the spread of an infection in a host population (Dunn *et al.* 1993*b*; Rigaud and Juchault, unpublished results).

Several other factors have been put forward to explain the scattered distributions of CSF and their low frequency in infected populations. These factors are: the selection of resistance genes (Rigaud and Juchault 1992, 1993), abiotic factors (temperature, salinity, photoperiod) limiting the spread of CSF (Bulnheim 1978; Ginsburger-Vogel and Desportes 1979; Rigaud *et al.* 1991*b*), random loss of infection in small subpopulations (Rigaud *et al.* 1992), and even, in some cases, a too high efficiency of CSF (no males are produced thus driving populations with high infection rates to extinction) (Werren 1987; Juchault *et al.* 1993). A combination of all these phenomena would allow the maintenance of other sex-determining systems in the overall range of a given species. The evolution of sex-determining mechanisms described in *A. vulgare* would also explain, in part, differences in feminizing factor prevalence (Juchault *et al.* 1993). The recurrent and unidirectional evolution from *Wolbachia* to the W chromosome via the f factor may explain why *Wolbachia* are often found associated with the f factor in populations, why the f factor can increase at the expense of *Wolbachia* in these populations, and why the W chromosome can be maintained in populations influenced by f (for detailed arguments, see Juchault *et al.* 1992, 1993).

The most obvious effect of feminizing factors at the population level is on sex ratio. Whatever the species, infected females produce 80–90 per cent daughters on average. The population sex ratio is therefore likely to depend on the frequency of infected females in populations. This fits well with the data on *O. gammarellus*, where high symbiont prevalence correlates with more female-biased population sex ratios (Ginsburger-Vogel and Desportes 1979). However, this relationship is less clear in *G. duebeni*, due to environmental effects on sex determination (Adams *et al.* 1987). In *A. vulgare*, the proportion of males in the field ranged from 19.6 to 60.7 per cent. These sex ratios are questionable since they do not obviously reflect the primary sex ratio, but Juchault *et al.* (1993) showed that the sex ratio produced by isolated females approximates well the overall population sex ratio. Populations infected with *Wolbachia* are always female-biased (from 26.5 to 39 per cent males), and populations showing a 1 : 1 sex ratio are those containing the higher proportions of WZ females. When the prevalence of the f factor was high, sex ratios were variable, i.e. from 19.6 to 52.9 per cent males, reflecting the variability of expression in the f factor.

Is a population sex-ratio bias detrimental or beneficial to the host? Apart from the obvious handicap of the disappearance of males, the answer remains unclear. In a model based on the *O. gammarellus* system, Wildish (1971) proposed that female-biased sex ratios can be beneficial to the host, until an optimum permitted by the environment is reached. The female-biased sex ratio would provide an advantage for the reproductive potential of the host below this value. In other words, population sex ratios between 50 and 85 per cent females would provide an advantage for a colonizing species, but above this value the rarity of males becomes limiting. For a species reaching its optimal density in a given environment, female-biased sex ratios would lead to an overproduction of young and would be selected against. Observations on some insect species tend to confirm the influence of sex-ratio distorters on population dynamics (discussed by Werren and Beukeboom 1993). James and Jaenike (1990) proposed that a driving X chromosome inducing female-biased sex ratios enhances the reproductive potential of populations of *Drosophila testacea*. On the other hand, Owen (1973) found that extreme female-biased sex ratios in wild populations of the lepidopteran, *Acraea encedon*, can lead to low mating frequency in females.

The presence of feminizing epigenetic elements may also drive the evolutionary stable sex ratio of the infected host (i.e. the ESS sex ratio produced by uninfected individuals) (Werren 1987; Hatcher and Dunn 1995). The sex ratio is expected to evolve toward monogeny if CSF transmission is less than 100 per cent; infected individuals would tend to produce only daughters while uninfected ones would produce only sons (Werren 1987). Variation could exist according to the transmission and the feminizing rates of the CSF, but ESS sex ratio will always be biased towards the non-transmitting sex (Hatcher and Dunn 1995). These models could explain some of the observations in *G. duebeni* wild populations.

3.3.3 Disturbance of sex chromosome (heterosome) evolution

The heterogametic system of sex determination has evolved independently in many different taxa, in parallel with gonochory (production of male and female gametes in separated individuals). The sex chromosomes are suspected to have evolved from autosomes in hermaphroditic ancestors in which there was strict linkage of male and female sex factors with a given chromosomal pair. The evolution of a number of serial phenomena is then necessary: failure of recombination between X and Y chromosomes through much of their length, genetic inertness of a large part of the Y chromosome, dosage compensation following this loss of functional genes, and finally the accumulation of repeated sequences on the Y chromosome (reviewed by Bull 1983; Charlesworth 1991).

In crustaceans, heterogametic sex determination seems to be the general rule, with several variants (Ginsburger-Vogel and Charniaux-Cotton 1982). However, in most cases, heterochromosomes are poorly differentiated morphologically (Lécher *et al.* 1995; Thiriot-Quiévreux and Cuzin-Roudy 1995). In the sub-order Isopoda, and more particularly in Oniscidea (woodlice), the heterogamety can only be distinguished indirectly, because of this morphological similarity (reviewed by Juchault and Rigaud 1995). Frequent crossing-over between sex chromosomes is known in the fresh-water Asellota, *Asellus aquaticus* (Vitagliano-Tadini 1963 [cited in Ginsburger-Vogel and Charniaux-Cotton 1982]). Moreover, the two heterogametic systems (XX/XY and XZ/ZZ) coexist in woodlice, sometimes within the same genus, and even within the same species (Table 3.1). Owing to these characteristics, one could assume that the evolution of heterogamety in Isopoda is at an incipient stage (Charlesworth 1991). It can be argued that the perturbation of sex determination by feminizing sex-ratio distorters can prevent the evolution of sex chromosome differentiation, i.e. can maintain this evolution at an early stage.

We have already noted that most species in the Oniscidea group are infected by *Wolbachia*, and that some of these microorganisms produce a feminizing effect (section 3.2.1). Thus, in this group, the sex determination is probably highly perturbed by CSF. By taking into account several possible ways to resolve the intragenomic conflicts induced by the presence of CSF and the evolutionary consequences of CSF on sex determination (section 3.3.1), it is

Table 3.1 Heterogametic types and heteromorphy in isopod crustaceans (from Juchault and Rigaud 1995)

Sub-order	Heterogamety type		
	♂ XO / ♀ XX	♂ XY / ♀ XX	♂ ZZ / ♀ WZ
Asellota		Asellus aquaticus**	Jaera marina (5 sub-sp.)***
Flabellifera	Tecticeps japonicus***		Dynamene bidentata*
Valvifera			Idotea balthica
Oniscidea		Porcellio dilatatus dilatatus*	Porcellio dilatatus petiti*
			Porcellio rathkei**
			Porcellio laevis**
		Armadilldium nasatum*	Armadillidium vulgare*
		Helleris brevicornis	Eluma purpurascens
			Oniscus asellus

Symbols indicate species with extreme (***), very slight (**) and no (*) heteromorphism of sex chromosomes. No symbol: no data about heteromorphism (in Oniscidea, heteromorphy is likely to be absent, because WW or YY individuals are viables and fertiles). The extreme heteromorphisms are always due to major events (lack of one sex chromosome or translocations).

possible that the chromosomal sex factors are not exclusively linked to a given chromosome pair. The most obvious example is given by the autosomal masculinizing gene, *M*, which became the male sex determinant in the presence of the f factor (Rigaud and Juchault 1993). In this case, a system similar to male heterogamety can be selected, but the heteromorphic chromosome pair would become the autosomal pair carrying the *M* gene (Box 3.3). Here, a male heterogamety can be seen as a by-product of the intragenomic conflict in a species with an ancestral female heterogamety. This would suggest that epigenetic sex factors can repeatedly change the location of sex-determining genes on their host chromosomes. It is important to note that this system possesses a pattern of epistasis (dominance across loci): *M* is epistatic to female sex factors (f and the sex factor carried by the W chromosome), which are in turn epistatic to the male sex factor carried by the Z chromosome (Table 3.2). The interesting point is that a similar pattern exists in the woodlouse species *Porcellio dilatatus*, where the two heterogametic systems coexist (XX/XY and WZ/ZZ). In this species, Y is epistatic to W and X, which are epistatic to Z (Legrand *et al.* 1980; Table 3.2). This hierarchy of dominance in *Porcellio* could be due either to more or less dominant genes at the same locus, but also at different loci, as is the case with the *M* gene. The origin of this double heterogamety system is unknown in *Porcellio*, but, owing to the similarities

Table 3.2 Hierachy of dominance (epistasis) of the different sex factors in two woodlouse species

Species	Sex factors and the associated phenotype	Epistasis pattern
Armadillidium vulgare	Z+Z = ♂	*M*>W=f>Z
	W+Z = ♀	
	W+W = ♀	
	f+Z = ♀	
	W+M = ♂	
	f+M = ♂	
Porcellio dilatatus	Z+Z = ♂	Y>W=X>Z
	W+Z = ♀	
	X+Z = ♀	
	X+X = ♀	
	W+Y = ♂	
	X+Y = ♂	
	Y+Y = ♂	

To simplify, each sex factor is denoted by the same letter as the chromosome that it is carried by ('Z' denotes 'the sex factor carried by the Z chromosome'), with the exception of *M* and f. 'X>X' must be read 'Y is epistatic to X'. It must also be pointed out that WW and XX individuals are females and that YY individuals are males.

between this system and the system involving the *M* gene, we can suggest that intragenomic conflicts have led to this situation.

The repetition of all these changes (changes of heterogamety system, changes in the sex factor location, changes in the nature of the sex factors [i.e. inhibitors, genes reducing bacterial transmission, etc.]) could have prevented sex chromosome differentiation and evolution, because of the instability of linkage between the sex factors and a given chromosomal pair. This instability could have prevented the lack of recombination between sex chromosomes because sex determination could be linked with other vital functions on these chromosomes. Highly differentiated sex chromosomes (heteromorphy) might even be deleterious in species with XY/XX heterogamety systems influenced by CSF. This is because a CSF always tends to eliminate the heterochromosome carrying the native female sex determinant (Taylor 1990). Whatever the heterogametic system involved, the infected individuals will tend to be homozygous for their male factors (i.e. infected individuals will be ZZ or YY at equilibrium). In cases of high heteromorphy, the ZZ genotypes are generally viable, but it is common to find that YY individuals are lethal or at least sterile (Bull 1983). Therefore, heteromorphy is expected to be selected against in cases of CSF-induced perturbation of sex determination (Taylor 1990). The only cases where YY individuals are viable is when heteromorphy is small or does not exist (for example in crustaceans: Juchault and Legrand 1979; Ginsburger-Vogel and Magniette-Mergault 1981*a*).

These arguments have not been rigorously tested empirically, but all the convergent data tend to indicate that CSFs could have prevented heterochromosome differentiation in a number of species. A possible way to test this hypothesis would be to investigate CSF efficiencies in the presence of various genotypes in a species with a double heterogametic system.

Acknowledgements

I wish to thank Alison Dunn, Ary Hoffmann, Pierre Juchault, Scott O'Neill and Jack Werren for having read an earlier version of the manuscript, and for their helpful suggestions and comments. I particularly thank Pierre Juchault for patiently guiding me in the labyrinth of crustacean sex determination.

4 *Wolbachia*-induced parthenogenesis

Richard Stouthamer

4.1 Introduction

The relative rareness of parthenogenetic reproduction in a predominantly sexually reproducing world has made it an intensively studied topic. Whereas the early work centred on the distribution of parthenogenetically reproducing forms (Phillips 1903; Winkler 1920; Vandel 1928), later interest focused on the cytogenetic processes that allow unfertilized eggs to develop into complete organisms (White 1970). In the mid-sixties, an emphasis on the evolution of sexual reproduction caused the emergence of a very active field of mainly theoretical studies into the immediate advantages of sexual reproduction compared to parthenogenetic reproduction (Williams 1966, 1975; Maynard Smith 1978). Theoretical work on the evolution of sex has again led to a review of parthenogenesis in order to find patterns against which some of the existing theories could be tested (Bell 1982). However, experimental verification of many of the theories on the evolution of sex remain difficult and rare.

The topic of this chapter is microbe-induced parthenogenesis, a recently discovered mechanism for parthenogenetic reproduction (Stouthamer *et al.* 1990*a*). It is still unclear how widely distributed this type of parthenogenesis is in animals. It has been verified in approximately 30 species of parasitic wasps, but a much wider distribution of this phenomenon is expected in insects and possibly other organisms (Luck *et al.* 1992). The discovery of bacteria-induced parthenogenesis was a logical consequence of extensive information that has been collected on the life history of parthenogenetic forms and the advances in sex-ratio theory by Hamilton (1967, 1979). Much of the life-history information has been collected by scientists working in the field of biological control of insect pests. The phenomenon of parthenogenetic reproduction has attracted their attention because completely parthenogenetic forms of parasitoids are considered desirable for use in biological control (Timberlake and Clausen 1924). Many of the early workers on biological control associated with the

University of California in Riverside, i.e. Clausen, Flanders and DeBach, have studied aspects of parthenogenesis (Timberlake and Clausen 1924; Clausen 1940; Flanders 1944, 1945, 1950, 1965; DeBach 1969; Rössler and DeBach 1972, 1973). Flanders (1965) was very close to discovering that parthenogenesis in some Hymenoptera can be reverted to sexual reproduction by temperature treatment of the wasps. Such temperature treatments have adverse effects on the bacteria associated with parthenogenesis (Stouthamer *et al.* 1990*a*).

The theoretical developments that have made the idea of microbial involvement in parthenogenetic reproduction likely are the studies by Hamilton on extraordinary sex ratios (1967, 1979) and work on the conflict in evolutionary interest between plasmagenes (i.e. genes located on heritable particles in the cytoplasm) and nuclear genes (i.e. genes located on the nuclear chromosomes) (Cosmides and Tooby 1981). These studies suggested that in sexual species plasmagenes should favour 100 per cent female progeny while nuclear genes should favour some other sex ratio always involving males. Many cytoplasmically inherited factors have been known to lead to female-biased sex ratios, but the extreme case of 100 per cent female offspring has never been considered to be caused by such factors because of the perceived need for fertilization. The involvement of cytoplasmic factors in causing parthenogenesis disqualifies such organisms to a large extent for experimental studies on the evolution of sex, which centres exclusively around the evolutionary interests of nuclear genes. In microbe-associated parthenogenesis the evolutionary interests of nuclear genes are, in most cases, subordinate to those of the plasmagenes.

This chapter reviews the work done on microbe-associated parthenogenesis, and it also indicates where large gaps exist in our understanding.

4.2 Terminology

Both males and females in most animal species are diploid, i.e. result from a fertilized egg. Eggs in these species generally only develop if a sperm has penetrated the egg and if the diploid number of chromosomes has been restored. Egg penetration is needed to initiate the developmental process. Even if no fusion of the sperm nucleus and egg nucleus has taken place, such eggs typically go through a number of divisions before the embryo dies. In such diplodiploid species, the transition from a normal sexual mode of reproduction to a completely parthenogenetic mode requires two modifications: first the egg needs to be able to develop without the stimulus provided by the sperm penetration and, secondly, the number of chromosomes needs to be restored to the diploid number. This diploidization can take place by a fusion of meiotic products of the egg or by a modification of meiosis in such a way that the diploid number of chromosomes is retained.

In haplodiploid species the normal mode of reproduction is also a form of parthenogenesis; males arise from unfertilized haploid eggs and females from fertilized, and therefore diploid, eggs. This mode of reproduction, which is a mixture of parthenogenesis (males) and sexual reproduction (females), is called arrhenotoky. Purely parthenogenetic reproduction, i.e. all offspring being produced from unfertilized eggs, is also found in haplodiploid species. Two forms of pure parthenogenesis are distinguished: deuterotoky and thelytoky (Winkler 1920). In deuterotoky both males and females emerge from unfertilized eggs, while in thelytoky all unfertilized eggs give rise to females. For the development of eggs in haplodiploid species the stimulus initiating the developmental process appears to be the deformation of the egg during oviposition. Therefore unfertilized haploid eggs develop in these species. Compared to diplodiploid species, the haplodiploid species have one less barrier to overcome in the transition from a sexual form of reproduction (arrhenotoky) to complete parthenogenesis (thelytoky).

Many terms are used to describe the ability of females to produce daughters without mating. In Hymenoptera such forms have been called uniparental, unisexual, or asexual. The more correct terms are: deuterotoky for taxa where unfertilized eggs give rise to males and females, and thelytoky where unfertilized eggs give rise to only females. This terminology became problematic when thelytokous females raised at high temperatures gave rise to offspring of both sexes whereas those raised at lower temperatures produced only daughters. Technically such situations call for the term thelytoky at low temperatures and deuterotoky at high temperatures. This problem has been solved pragmatically by DeBach (1969) who called those species that produced less than 5 per cent males thelytokous and those that produced a higher percentage of males deuterotokous. Recently it has been suggested by Luck *et al.* (1992) that the term deuterotoky be abolished altogether and that thelytoky be used for those situations in which virgin females are capable of producing some daughters from unfertilized eggs. However, it has again become more complicated since we now know that in some cases thelytokous females can produce daughters both from fertilized and unfertilized eggs. Here the term thelytoky will be used for the mode of reproduction in which daughters can emerge from unfertilized eggs. Arrhenotoky will be used for the mode of reproduction where males arise from unfertilized eggs and females from fertilized eggs.

4.3 Evidence for involvement of *Wolbachia* in thelytoky

In several thelytokous parasitic wasp populations that were reared or monitored for biological control projects, males appeared occasionally during the summer or when the wasps were reared under elevated temperatures (Perkins 1905; Smith 1941; Flanders 1945). Systematic studies of temperature influences on the gender of the offspring of thelytokous females were reported first by Wilson and Woolcock (1960*a*,*b*) for the thelytokous parasitoid *Ooencyrtus submetallicus*. The temperature to which the female parent was exposed during development and adult life determined the proportion of males in her offspring. Temperatures above 29.5 °C caused only males to be produced. Similar results were found later in a number of other thelytokous species such as *Pauridia peregrina* (Flanders 1965), *Trichogramma* sp. (Bowen and Stern 1966), *T. oleae* (Jardak *et al.* 1979), *T. cordubensis* (Cabello and Vargas 1985) and *Ooencyrtus fecundus* (Laraichi 1978). The threshold temperature for male production ranged from above 28 °C, for *Trichogramma* spp., to temperatures higher than 30 °C for *O. fecundus* and *P. peregrina*. For many other thelytokous wasps, temperature effects on the sex ratio of their offspring were reported (Table 4.1).

Males produced by thelytokous females that had been treated with elevated rearing temperatures appeared in a number of cases to be able to father offspring with arrhenotokous conspecifics (Rössler and DeBach 1972,1973; Orphanides and Gonzalez 1970) although in many other cases similar crosses did not succeed (Table 4.1).

Generally the ability to produce daughters from unfertilized eggs was considered a nuclear chromosomal trait. Evidence for a simple Mendelian gene for thelytoky exists in *Apis mellifera* (Kerr 1962, but see Rüttner 1988). To determine whether thelytoky in parasitoid Hymenoptera was a simple Mendelian trait, Stouthamer *et al.* (1990*b*) used *Trichogramma pretiosum* males from temperature-treated thelytokous lines to introgress the nuclear genome of the thelytokous line into a conspecific arrhenotokous line. At every generation of introgression the mode of reproduction of the females was tested by allowing virgin females to produce offspring at normal rearing temperatures. Even after nine generations of introgression, when 99.6 per cent of the genome of the tested females should have come from the thelytokous line, these females still produced only male offspring. Thus thelytoky was not inherited as a simple Mendelian trait.

Extrachromosomal inheritance of thelytoky was subsequently studied by administering antibiotics to thelytokous females (Stouthamer *et al.* 1990*a*). This led to the appearance of males in the offspring of treated females and not in those of control females. Thus antibiotics appeared to kill a microorganism causing thelytoky. Several antibiotics were tested and only three were effective in causing the production of male offspring: tetracycline hydrochloride, sulphamethoxazole and rifampin. In those species where males were produced following antibiotic treatment, temperature treatment had the same effect. In other thelytokous species neither antibiotic nor temperature treatment led to the production of male offspring (Stouthamer *et al.* 1990*a*). Subsequently, the

Table 4.1 Cases of thelytokous reproduction in which evidence exists of *Wolbachia* involvement

Taxon	h[a]	a[a]	w[a]	p[b]	c[c]	References
Tenthredinoidea						
Pristiphora erichsonii	+	?	?	?	?	Smith (1955)
Chalcidoidea						
Pteromalidae						
Muscidifurax uniraptor	+	+	+	f	–	Legner (1985*a*, *b*) Stouthamer *et al.* (1993, 1994)
Aphelinidae						
Aphelinus asynchus	+	?	?	?	?	Schlinger and Hall (1959)
Aphytis diaspidis	?	?	+	?	?	Zchori-Fein *et al.* (1994, 1995)
A. lignanensis	?	+	+	?	+[d]	Zchori-Fein *et al.* (1994, 1995)
A. mytilaspidis	?	?	?	m	+	Rössler and DeBach (1973)
A. yanonensis	?	+	+	?	?	H. Nadel (personal communicatiom), Werren *et al.* (1995)
Encarsia formosa	?	+	+	f	–	Zchori-Fein *et al.* (1992), Van Meer *et al.* (1995), Werren *et al.* (1995)
Signiforidae						
Signiphora borinquensis	+	?	?	?	+	Quezada *et al.* (1973)
Encyrtidae						
Aponanagyrus diversicornis	+	+	+	f	+[e]	Pijls *et al.* (1996)
Pauridia peregrina	+	?	?	f	+	Flanders (1965)
Ooencyrtus submetallicus	+	?	?	?	–	Wilson and Woolcock (1960*a*, *b*), Wilson (1962)
O. fecundus	+	?	?	?	?	Laraichi (1978)
Plagiomerus diaspidis	+	?	?	?	–	Gordh and Lacey (1976)
Trechnites psyllae	?	+	?	?	?	Unruh (personal communication)
Habrolepis rouxi	+	?	?	?	?	Flanders (1965)
Trichogrammatidae						
Trichogramma brevicapillum	+	+	+	m	+	Stouthamer *et al.* (1990*a*, *b*), Werren *et al.* (1995)
T. chilonis	+	+	+	m	+	Stouthamer *et al.* (1990*a*, *b*), Chen *et al.* (1992), Schilthuizen (peronal communication)
T. cordubensis	+	+	+	f	+	Cabello and Vargas (1985), Stouthamer *et al.* (1990*b*, 1993), Silva and Stouthamer (1996)
T. deion	+	+	+	m	+	Stouthamer *et al.* (1990*a*, *b*, 1993)
T. embryophagum	+	+	?	?	+	Birova (1970, Stouthamer *et al.* (1990*b*)
T. evanescens (*rhenana*)	+	+	?	?	+	Stouthamer *et al.* (1990*b*)
T. nr deion (*KK*)	?	+	+	m	+	Stouthamer and Kazmer (1994), Schilthuizen (personal communication)
T. oleae	+	+	+	?	+	Louis *et al.* (1993)

T. platneri	+	+	?	m	+	Stouthamer *et al.* (1990*a*)
T. pretiosum	+	+	+	m	+	Orphanides and Gonzalez (1970), Stouthamer *et al.* (1990*a*, *b*)
T. near sibericum	+	?	+	?	?	Schilthuizen and Stouthamer (personal communication)
T. sp.	+	?	?	?	?	Bowen and Stern (1966)
T. telengai	+	?	?	?	?	Sorakina (1987)
Cynipoidea						
Hexacola sp. *near websteri*	+	?	?	?	?	Eskafi and Legner (1974)
Leptopilina australia	?	?	+	?	?	Werren *et al.* 1995*a*)
L. clavipes	?	?	+	f	–	Werren *et al.* 1995*a*)
Diplolepis rosae	?	?	+	f	–	Stille and Dävring (1980), van Meer *et al.* (1995)

The evidence is classified as males following heat treatment (h), males following antibiotic treatment (a), molecular evidence for *Wolbachia* presence (w). In addition information is given if the thelytokous forms are found in populations where thelytoky is fixed in the population or if it occurs mixed with arrhenotoky (p), and if the males and females are capable of successful copulations (c).

[a] +, Evidence exists; ?, information not available.

[b] f, Thelytoky fixed in population; m, thelytoky and arrhenotoky occur in populations; ?, information not available.

[c] Copulations are successful (+) or not (–).

[d] Mating and sperm transfer take place, but no successful fertilization of eggs.

[e] Mating of males of thelytokous line is succssful with closely related arrhenotokous females, but not with thelytokous females.

presence of bacteria in eggs of thelytokous females was demonstrated by Stouthamer and Werren (1993) using lacmoid staining. Microorganisms were not present in the eggs of lines that had been 'cured' by antibiotic or temperature treatment, nor in arrhenotokous field-collected lines or in thelytokous lines that did not respond to antibiotic treatment. Therefore there was a strong correlation between the presence of microorganisms and the presence of revertible thelytoky in *Trichogramma* species. The microorganisms in thelytokous *Trichogramma* species have subsequently been identified as *Wolbachia* (Rousset *et al.* 1992*a*; Stouthamer *et al.* 1993). *Wolbachia* has also been found in other thelytokous parasitic wasp species (Table 4.1). The final proof of the causal involvement of *Wolbachia* in thelytoky is awaiting our ability to make an arrhenotokous line thelytokous by infection with *Wolbachia*.

4.4 Phylogeny of *Wolbachia* associated with parthenogenesis

Several effects are associated with *Wolbachia* infections, i.e. incompatibility (Chapter 2), feminization (Chapter 3) and parthenogenesis. The *Wolbachia* associated with parthenogenesis do not form a monophyletic group within the phylogeny of *Wolbachia*. In both the phylogenies of *Wolbachia* strains based on the 16S RNA gene (Rousset *et al.* 1992*a*; Stouthamer *et al.* 1993) and the *ftsZ* gene (Werren *et al.* 1995*a*), the parthenogenesis-associated *Wolbachia* are intermixed with *Wolbachia* causing other effects, such as cytoplasmic incompatibility. Several hypotheses can explain such polyphyly:

1. Multiple evolution of induction of parthenogenesis. If this is the case, the transition between incompatibility and parthenogenesis must be an easy evolutionary step because many independent transitions need to be hypothesized to explain the current distribution of the parthenogenesis trait in the *Wolbachia* phylogeny.
2. Host effects, i.e. the same *Wolbachia* causes parthenogenesis in one host and incompatibility in other hosts. It is well established in the *Wolbachia* strains causing incompatibility that the host can have a strong influence on the phenotypic expression of a particular *Wolbachia*. For instance, the *Wolbachia* infections of *D. simulans* cause reduced levels of incompatibility in *D. melanogaster* (Boyle *et al.* 1993). The test of this hypothesis awaits our ability to establish an infection with the parthenogenesis *Wolbachia* in a host normally associated with cytoplasmic incompatibility.
3. *Wolbachia* genes associated with the induction of parthenogenesis or incompatibility are not linked to the genes that the phylogeny is based on and DNA exchange takes place between different *Wolbachia*. Such DNA transmission may take place through plasmids or other virus-like particles. The phage-like particles observed in *Wolbachia* in mosquitoes (Wright *et al.* 1978) could be examples of vehicles for such DNA transfer. But at this time there is no evidence for the general presence of plasmids or viruses in *Wolbachia*.

4.5 Evidence for horizontal transmission in parthenogenesis *Wolbachia*

Few detailed studies exist on the distribution of parthenogenesis *Wolbachia* in closely related hosts. Only in the genus *Trichogramma* have a substantial number of parthenogenetic forms of several species been studied. These studies (Stouthamer *et al.* 1993; Werren *et al.* 1995*a*; Schilthuizen and Stouthamer, in preparation) show that the *Wolbachia* in *Trichogramma* form a monophyletic group restricted to this host genus. But within different populations of the same species, different types of *Wolbachia* can be found, and no correlation exists between the phylogeny of the *Wolbachia* and that of the *Trichogramma* species (Schilthuizen and Stouthamer 1997). Apparently, horizontal transmission of the *Trichogramma Wolbachia* takes place between *Trichogramma* species. The frequency of horizontal transmission may be studied by comparing the parthenogenesis *Wolbachia* sharing hosts in populations of different *Trichogramma* species.

The monophyly of the parthenogenesis *Wolbachia* associated with the genus *Trichogramma* contrasts with the presence of unrelated incompatibility *Wolbachia* in the genus *Drosophila*. To determine if such monophylly is a feature setting the parthenogenesis *Wolbachia* apart from the incompatibility *Wolbachia*, more studies are needed on the phylogeny of the *Wolbachia* from host species of a particular genus. Taxa that could be used for such studies include the genera *Encarsia* and *Aphytis*, since in both these taxa a high incidence of parthenogenesis is found. Initial information on the phylogenetic position of the *Wolbachia* in *Aphytis* (Werren *et al.* 1995*a*; Zchori-Fein *et al.* 1995) show that this group may not be very useful because it appears to fall in the group of *Wolbachia* with little phylogenetic divergence, i.e. group 2 (Breeuwer *et al.* 1992; Stouthamer *et al.* 1993) or type A (Werren *et al.* 1995*a*).

4.6 Cytogenetics of *Wolbachia*-associated thelytoky

Many cytogenetic processes are known that can result in a diploid offspring from an unfertilized egg. These processes can be divided in essentially two groups:

1. Meiotic modifications: modifications take place before or during meiosis in such a way that the number of chromosomes during the meiotic process is not reduced to the haploid number, or this number is restored by the fusion of two of the meiotic products.
2. Postmeiotic modification: the diploid number of chromosomes is restored after a normal meiosis has resulted in one pronucleus, for instance by the fusion of haploid mitotic products.

Most cases of parthenogenetic development known in insects belong to the first group (Suomalainen *et al.* 1987). Numerous cytogenetic processes have been recorded in thelytokous Hymenoptera (Crozier 1975); however, only two publications describe in detail the cytogenetics of wasps now known to be infected with *Wolbachia* (Stille and Dävring 1980; Stouthamer and Kazmer 1994). Stouthamer and Kazmer (1994) studied the cytogenetic processes by observing chromosome behaviour and by recording the genetic consequences of the mechanism for the restoration of diploidy. The chromosome behaviour studies showed that meiotic processes taking place after an unfertilized egg has been laid, i.e. from anaphase I of meiosis onward, are the same as those in eggs from uninfected (arrhenotokous) females (Fig. 4.1). Thus the eggs progress to the stage where there is a single pronucleus and two polar bodies. At this stage, in an arrhenotokous egg the development is arrested for a certain period, and if no sperm arrives the haploid pronucleus will go into the first mitotic division. During this division the number of chromosomes is doubled and during the anaphase of the first mitotic division each haploid

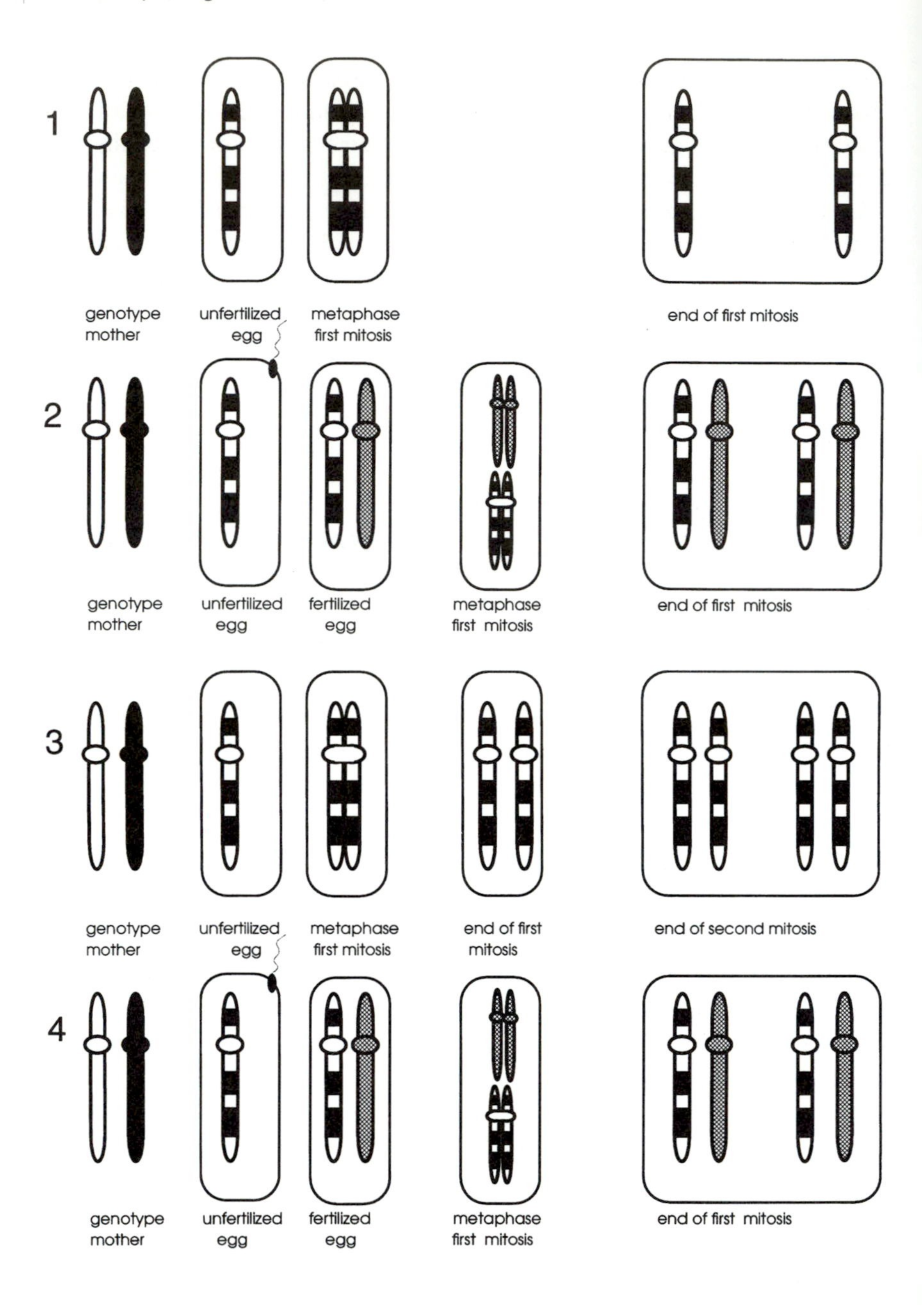

Figure 4.1 Diagram showing cytogenetic mechanisms (from anaphase I of meiosis to the second mitotic division) in eggs differing in their infection with parthenogenesis *Wolbachia* and whether they are fertilized or not. For simplicity only one of the sets of chromosomes is shown per haploid genome. (1) Uninfected unfertilized egg goes through a number of mitotic divisions conserving the haploid number of chromosomes per nucleus. Such an egg grows out to be a normal haploid male (i.e. has one set of chromosomes per nucleus). (2) Uninfected fertilized egg. By fusion of the sperm nucleus with the egg pronucleus, the diploid number is restored. Such an egg grows out to be a normal diploid female (i.e. has two sets of chromosomes

set of chromosomes is pulled to a different pole. In this way two nuclei are formed each with a haploid number of chromosomes. Similar mitotic divisions follow, leading to a large number of haploid nuclei which will eventually result in the formation of males.

In the eggs of infected females the meiosis also progresses to the stage of a single haploid pronucleus. The restoration of the diploid number takes place in the first mitotic division (Fig. 4.1). During the anaphase of the first mitotic division, the two identical sets of chromosomes do not separate and the net result of the first mitotic cycle is a single nucleus containing two copies of the same set of chromosomes. This stage is followed by normal mitotic divisions, eventually resulting in a female offspring that is completely homozygous at all loci. This cytogenetic mechanism is known as gamete duplication. The cytogenetic observations on two thelytokous *Trichogramma* forms have been verified using allozyme markers. In some thelytokous *Trichogramma* forms, as well as in a few other thelytokous parasitoids, infected females mate and use the sperm from conspecific arrhenotokous males to fertilize their eggs. Such fertilized eggs become infected females whose genome consist of one set of chromosomes from their father and one from their mother. By crossing lines that differed in various electrophoretic markers, Stouthamer and Kazmer (1994) showed that the thelytokous females were indeed capable of using sperm from arrhenotokous lines. Such mated infected females fertilized their eggs in approximately the same proportions as uninfected females. These heterozygous infected females were allowed to produce offspring as virgins and the segregation of the allozyme markers was studied in the female offspring. All offspring tested were homozygous for the allozymes used as marker. No heterozygous offspring were found in the 650 offspring tested. This result confirms the observations on the chromosome behaviour. In addition, crossing-over of linked markers in the offspring of these females indicated that crossing-over takes place during oogenesis in the thelytokous females. This makes it likely that the meiosis in infected and uninfected females is identical and that the only change taking place in the infected eggs is the doubling of the number of chromosomes in the first mitotic division if the eggs remain unfertilized. If the infected eggs become fertilized, the diploidization by the sperm supersedes the influence *Wolbachia* may have.

The chromosome behaviour was also studied in detail in the thelytokous gall wasp *Diplolepis rosae* (Stille and Dävring 1980). The mechanism of restoration of diploidy was essentially the same as that described by Stouthamer and

---

per nucleus). (3) Infected unfertilized egg, *Wolbachia* causes an aborted first mitotic division, resulting in the two sets of chromosomes in one nucleus (i.e. diploid). The next mitotic division is normal, resulting in two diploid nuclei. Such eggs grow out to be completely homozygous diploid infected females. (4) Infected fertilized egg, the sperm nucleus fuses with the pronucleus and the infection with *Wolbachia* does not interfere with the mitotic divisions. Such eggs grow out to be biparental infected females.

Kazmer (1994) for *Trichogramma.* The rare males of the gall wasp could not be crossed with conspecific females and therefore no segregation of markers could be used to confirm the gamete duplication that was observed in the eggs. However, Stille (1985) determined that all wasps collected from a number of locales throughout northern Europe were completely homozygous for all markers studied. Later van Meer *et al.* (1995) showed that *D. rosae* is also infected with *Wolbachia.* It has not yet been shown that the thelytoky of *D. rosae* can be cured by killing the *Wolbachia* symbiont. Males are produced occasionally and at higher frequencies in certain areas, but no clear correlation seems to exists with temperature (Callan 1940). Finally, the mechanism of restoration of diploidy in the thelytokous *Muscidifurax uniraptor* has been reported to be endomitosis by Legner (1985*a*). In this species *Wolbachia* has been found and infected females produce male offspring following antibiotic treatment (Stouthamer *et al.* 1994). In all cases where the cytogenetics of *Wolbachia*-associated parthenogenesis has been studied, the mechanism of restoring diploidy is gamete duplication.

In one other species (*Aphytis mytilaspidis*), where we suspect *Wolbachia* to be the cause of thelytoky, the chromosome behaviour has been studied during meiosis (Rössler and DeBach 1973). *Wolbachia* involvement is suspected because the production of males is sensitive to temperature treatment and *Wolbachia* has been found in other members of this genus (Zchori-Fein *et al.* 1995). Rössler and DeBach (1973) have shown that meiosis appears to be normal but they were not able to resolve the chromosome behaviour completely during the restoration of diploidy. As in *Trichogramma* spp., the thelytokous *Aphytis* was also capable of mating and would use the sperm of the arrhenotokous line to fertilize eggs. The female offspring of virgin heterozygous females consisted mainly of females homozygous for the markers, but a number of heterozygous offspring has also been found. Thus gamete duplication cannot be the sole form of restoration of diploidy. Fusion between the pronucleus and a polar body has taken place at least in some eggs. Rössler and DeBach concluded that the cytogenetic mechanism is most likely terminal fusion.

Thus it appears that *Wolbachia* infection generally results in gamete duplication as the mechanism for the restoration of diploidy. However, other mechanisms cannot be excluded until more detailed work is done. Gamete duplication appears to be a relatively rare cytogenetic process in insects (Suomalainen *et al.* 1987), possibly because recessive lethal and deleterious genes are exposed immediately, while other cytogenetic mechanisms that repress meiosis or are based on a fusion between a polar body and the pronucleus can maintain heterozygosity for a longer period. Another case of gamete duplication in haplodiploids has been reported from *Trialeurodes vaporariorum* (Thomson 1927), but it is unknown whether *Wolbachia* is involved. The cytological effects of *Wolbachia* are now understood, but how these prokaryotes influence the biochemical or mechanical processes during the first mitotic anaphase remains to be studied.

4.7 Gynandromorphs

The temperature to which infected females are exposed during their larval and pupal development influences the gender of the offspring they produce. Only daughters are produced if the temperature at which the mother develops is below 26 °C, while only male offspring are produced at temperatures above 30 °C. These temperatures apply to a number of species (*Trichogramma* spp. and *Ooencyrtus submetallicus*), while in *Ooencyrtus fecundus* the change from female production to male production lies between 30 and 35 °C. It is not yet known if this temperature effect is caused by a physiological change in the wasp or by a change in the *Wolbachia*, or both. Not only does the mother's developmental temperature affect the sex ratio of her offspring but also the temperature at which she is kept after oviposition (Laraichi 1978). The transition from producing only daughters to only sons can be very abrupt, for instance in *O. submetallicus* a rearing temperature of 29.2 ± 0.3 °C resulted in only daughters, whereas a rearing temperature of 29.4 ± 0.3 °C led to the production of males only (Wilson and Woolcock 1960*b*).

At intermediate temperatures (28–30 °C) between the all-male and all-female progeny, many thelytokous wasps produce females, males and gynandromorphs, i.e. individuals in which some of the tissues are female and some are male (Wilson and Woolcock 1960*b*; Bowen and Stern 1966; Cabello and Vargas 1985). The gender of the tissue is most likely determined by the level of ploidy of the blastoderm cell from which the tissue is derived. A possible scenario is that the effect of *Wolbachia* on gamete duplication is somewhat repressed in eggs from mothers exposed to these temperatures and that the gamete duplication does not take place in the first mitotic division but in one or more nuclei of a later division (Cooper 1959). Thus one or several nuclei will be diploid and the rest will be haploid. The level of male tissues varies with the temperature at which the mother of the gynandromorph has been reared; the closer to the all-male boundary the larger the proportion of male tissue. Such a trend could be caused when a gamete duplication takes place in one nucleus in a later mitotic division. When the gamete duplication takes place in one of the two nuclei of the second mitotic division, the proportion of cells in the embryo should be 1 female to 2 male, assuming that both diploid and haploid nuclei divide at a similar rate. This proportion is 1 female to 7 male in the third mitotic division and 1 to 14 in the fourth. Thus the timing of the gamete duplication may explain the reduction in fraction of female tissue with increasing temperature.

For this hypothesis to apply, the lowest frequency of male tissue should at least be two-thirds of all tissues, but this does not appear to be the case. In some wasps the only male tissue is found on one of the two, otherwise female, antennae. If this hypothesis is correct, we must either assume a differential division rate of the haploid and diploid cells or the diploid cells must preferentially be included in those participating in the formation of the blastoderm. Alternatively, the stage at which the gamete duplication takes place may be moved to a later mitotic division and the duplication should take place in not just one nucleus but in the majority of nuclei. Such a mechanism of duplication has been found in a cave cricket (Lamb and Willey 1989). If this duplication took place in random nuclei, any tissue would have a chance of either being male or female. However, there always appears to be a polarity to the male and female tissues found in these gynandromorphs. Individuals with little male tissue always have the male tissue on the head and female tissues for the rest of the body. Gynandromorphs formed at higher temperatures have the head male and the rest of the body female, and those formed at highest temperatures have male tissue on the abdomen as well. The reason for this polarity is not yet known, possibly the modification of the mitosis that results in a diploid nucleus always takes place in a certain area of the egg, which may be influenced by the *Wolbachia* densities or the concentration of some product produced by *Wolbachia*. In *Trichogramma* eggs, *Wolbachia* is concentrated in the posterior part of the egg, close to the germ-cell determinant, which will develop into the ovaries.

4.8 Fitness effects of *Wolbachia* infection on the parasitoid host

The influence of *Wolbachia* on life-history traits of its host can only be unambiguously determined when infected and uninfected genetically identical females are compared. Such a comparison can be made in those species where sexual lines can be established. Thus far this has only been possible in *Trichogramma* spp. In *Trichogramma* wasps a substantial negative influence of *Wolbachia* on the offspring production is found under laboratory conditions. If wasps have unlimited access to hosts, infected wasps in all cases produce fewer offspring than the wasps that are cured of their infection. In several cases, the number of daughters produced by infected mothers was even lower than that produced by the uninfected mothers. The thelytokous *Trichogramma* forms that have been tested have all been collected from mainly arrhenotokous field populations (Stouthamer *et al.* 1990*b*; Stouthamer and Luck 1993). It is clear that for the infection to remain in such field populations the infected mothers must at least produce similar numbers of infected daughters as uninfected mothers produce uninfected daughters. Therefore, these laboratory results cannot reflect what happens in the field. Stouthamer and Luck (1993) showed that if females are limited in the number of hosts they encounter during their life, as is probably the case in the field, the infected females produce more daughters than uninfected females and this may explain in part the prolonged coexistence of infected and uninfected conspecifics in the field.

In those species where infection has gone to fixation and cured lines cannot be established, it is not possible to measure directly the influence of the *Wolbachia* on offspring production because infected and uninfected females cannot be compared. The only way to test whether the infection has a negative effect on offspring production is to compare the offspring production of adult females fed honey with similar females fed antibiotics. Such tests were done with *Encarsia formosa* and *M. uniraptor*, both species of which all known populations consist of infected females and sexual lines cannot be established. In these species, no negative effect of the *Wolbachia* infection could be detected (Stouthamer *et al.* 1994) . Similar experiments done with *Trichogramma* show a significant negative effect of the infection (Horjus and Stouthamer 1995).

Explanations for the differences in effects on offspring production in the cases where the infection has gone to fixation and those where infected and uninfected females still coexist can be:

1. In the species where the infection has gone to fixation, the initial *Wolbachia* infection had little or no negative impact on the offspring production of the females. When the complete population is infected, subsequent selection favours those *Wolbachia*–host combinations that are the most fit. These can come about by either changes in the *Wolbachia* genome, the host genome, or by an interaction between the two genomes.
2. In populations where the *Wolbachia* infection has not gone to fixation, a conflict exists between the interest of the nuclear genome and that of the cytoplasmically inherited genome (i.e. *Wolbachia* and mitochondria). The optimal sex ratio for the cytoplasmic genome is 100 per cent daughters (Cosmides and Tooby 1981), while that of the nuclear genome favours the production of at least some males, as long as infected females mate. The higher the frequency of infection, the more reproductive fitness can be gained through males. Thus nuclear genes that would render infected females into at least partial male producers would spread through the population. The nuclear–cytoplasmic conflict may result in a response of the *Wolbachia*, for instance an elevation in the concentration of *Wolbachia* that could compensate for the effects of the suppressor genes. A side-effect of this elevated concentration could be a negative effect on the offspring production of the infected females. Suppressor genes against maternally inherited male-killing factors, most probably symbionts, have been found in *Drosophila prosaltans* (Cavalcanti *et al.* 1957) and against

the feminizing *Wolbachia* in *Armadillidium vulgare* (Rigaud and Juchault 1992).

Only indirect evidence exists for the presence of suppressor genes to parthenogenesis-inducing *Wolbachia*. First, some field-collected eggs of *T. nr. deion* from the Mojave desert gave rise to only females; when these virgin females were allowed to produce offspring some produced all-female offspring while others produced all-male offspring (Stouthamer, unpublished). This might happen if an infected mother heterozygous for a recessive suppressor has mated with a suppressor male. Her offspring from fertilized eggs should consist of a 1 : 1 ratio of infected and neo-uninfected females (i.e. those homozygous for the suppressor), and therefore produce only male offspring when reproducing as a virgin. A systematic search for suppressor genes is now under way in these populations. The second piece of evidence is even more indirect. Rössler and DeBach (1972) studied thelytoky in *Aphytis mytilaspidis*, a case of thelytoky that appears likely to be caused by *Wolbachia* infection, although this has not yet been proved. In crosses between thelytokous females and genetically marked males from an arrhenotokous line, they created hybrid females that could be recognized by visible markers. These hybrid females were allowed to produce offspring as virgins. Some produced only female offspring and others only male offspring. In crosses with males of one type, 23 out of 120 hybrid females produced only sons, while in the other cross 16 out of 23 hybrid females produced only sons. The genetic make-up of the father seems to influence the frequency of thelytoky in his offspring. Clearly, more systematic searches for suppressors are needed.

4.9 Dynamics of infection frequency in populations

The frequency of infection in populations studied thus far is bimodal. In most known cases of *Wolbachia*-associated thelytoky the infection has gone to fixation. An exception to this rule is in *Trichogramma* populations, where in most species a low frequency of infection occurs. The exceptional status of *Trichogramma* species may be caused by the sampling method with which many populations have been studied. Each female offspring from field-collected hosts has been the foundress of an isofemale line, allowing for the detection of rare thelytokous individuals (Pinto *et al.* 1991). There is indirect evidence that populations of other species may also be polymorphic for thelytoky.

In a number of cases, species collected for biological control of pests were thought to be arrhenotokous in their native range, but turned thelytokous during laboratory rearing or subsequent release for biological control (Hung *et al.* 1988). A possible explanation for this change in mode of reproduction is that the native populations consisted of a mixture of thelytokous and arrhenotokous individuals and that during laboratory rearing or subsequent release the thelytokous individuals outcompeted arrhenotokous ones (Stouthamer and Luck 1991; Stouthamer 1993).

4.9.1 Polymorphic populations

4.9.1.1 Polymorphism through inefficient transmission

The dynamics of the infection within mixed populations depends on the relative offspring production of the infected versus the uninfected females (w = total offspring production of an infected female divided by the total offspring of an uninfected female), the rate of vertical transmission (α = the frequency of infected daughters in the offspring of infected mothers), and the sex ratio produced by the uninfected fraction of the population (x = proportion of females in the progeny of uninfected females, i.e the fertilization rate of the eggs). If we assume that all infected and uninfected females mate and fertilize a fraction, x, of their eggs, it can be shown that the equilibrium fraction of infection (F) among females in the population is given by:

$$F = w\alpha - x/w\alpha + (1-\alpha)\,wx - x.$$

It is clear that if the vertical transmission is perfect, any equilibrium value can be maintained if the relative offspring production of the infected females equals the sex ratio produced by the uninfected females. These are neutral equilibria. As soon as the transmission of infected females becomes less than perfect ($\alpha < 1$), stable equilibria ($0 < F < 1$) may be attained when $w\alpha > x$. Generally, the transmission rate in laboratory populations of *Trichogramma* is high, greater than 90 per cent (Stouthamer, unpublished data). The higher the sex ratio (x) of the uninfected population, the higher w has to be for a particular equilibrium frequency to be reached. Thus a population with a lower x in the uninfecteds is easier to invade and a higher equilibrium can be attained given a constant value for w. When w is relatively small, only low equilibrium rates may be attained, but w must always be larger than x for the infected fraction to remain in the population. The relationship between the rate of infection among females and the relative offspring production (w) is given in Fig. 4.2. In this figure a fertilization rate (x) of 75 per cent is used, such a value is found in many *Trichogramma* populations.

In most *Trichogramma* populations that have been studied in some detail the infection frequency is less than 5 per cent (Fig. 4.2), and assuming a vertical transmission rate (α) of 0.9, such an equilibrium rate would be possible if $w = 0.95x/(0.855-0.005x)$. For a sex ratio of these wasps in the field of approximately 80 per cent females, this means that the infected females should produce at least 89.4 per cent of the total offspring produced by the

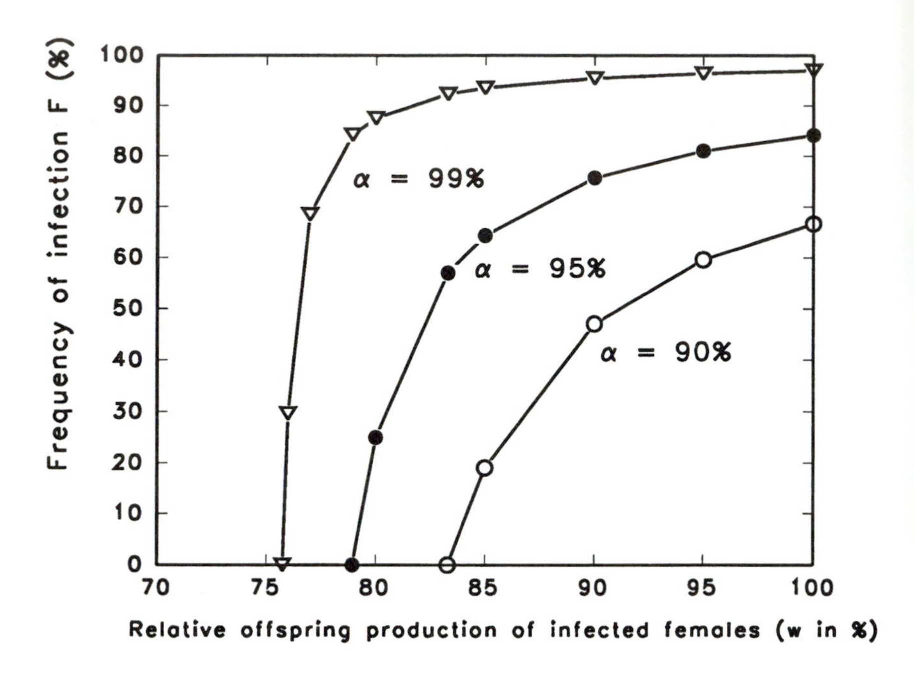

Figure 4.2 Relationship between the frequency of infection among females (F) as a function of the relative offspring production of infected females (w), when the fertilization rate in the population (x) equals 75 per cent, for different levels of transmission efficiency of the *Wolbachia* (α).

uninfected females. For values of α closer to 1, w should approach x. One of the reasons why the *Wolbachia* infections in *Trichogramma* may not have gone to fixation is that normally the uninfected females produce relatively high (i.e. female biased) sex ratios, requiring large values of w for a high equilibrium.

4.9.1.2 Polymorphism through suppression of *Wolbachia*

Once a certain fraction of the population has become infected with *Wolbachia*, a nuclear suppressor of the *Wolbachia* that induces infected females to produce uninfected offspring may spread through the population. The higher the infected fraction in the population, the larger the fitness advantage for such a suppressor. A necessary condition for the spread of the suppressor and the maintenance of the equilibrium is that w is larger than x, and if we assume a perfect transmission of the *Wolbachia*, the cost of being homozygote for the suppressor has to be approximately equal to the difference between w and x. Simulations (Stouthamer, unpublished) show that the presence of such a suppressor allows for very robust equilibrium values of infection of around 10–20 per cent of the females. The equilibrium is maintained because the negative effects of carrying the suppressor in homozygous form for females is balanced by the increased transmission through males mating with infected females. Thus in such a system the level of *Wolbachia* infection is kept in check by the fact that infected females still mate and are exposed to the effects of the suppressor gene.

4.9.1.3 Polymorphism through suppression of *Wolbachia* and inefficient transmission

A combination of both suppressor genes and inefficient transmission allows a stable equilibrium in infection rates to exist. If suppression forms an important component of keeping the level of infection low, such a population could still go to fixation for the infection. This can take place if a nuclear mutant arises within the infected fraction of the population that increases the mating reluctance of females. Mutant infected females do not mate and thus are no longer exposed to the suppressor gene. The fraction of the infected individuals carrying the mating reluctance mutation will increase over time and this mutant will go to fixation in the infected population. If there still is inefficient transmission (i.e. $\alpha < 1$) then the infected mutant females will produce mutant males from uninfected eggs. These males mate with the uninfected females and this results eventually in a decreased mating willingness in some of the uninfected females. Such a scenario causes a reduction in the average fertilization rate (x) and this again contributes to a higher frequency of the infection in the population, possibly leading to the fixation of *Wolbachia*.

4.9.2 Fixation of the infection in populations

Once infection rates have reached a high level in the population, other effects will cause a further decrease of uninfecteds in a population, even though the equation above would predict that a small fraction of the population would remain uninfected. When the frequency of infection among the females is high, males will be rare. Assuming that infected females can only produce uninfected daughters when they lay uninfected, fertilized eggs, the frequency of these uninfected females depends on the frequency of males in the population. When males are rare in the population because of a high infection rate, the relative shortage of males will result in mainly male offspring from uninfected eggs. Therefore at higher levels of infection, a further decay in the number of uninfected females is expected. Once very high levels of infection have been reached, all the uninfected individuals that are still in the population will be males resulting from unifected eggs produced by infected females through inefficient transmission.

The scenario described above indicates that it will be difficult for arrhenotoky to invade thelytokous populations because of the low frequency of males and the competition between infected and uninfected females for mates. The negative effects of such competition for the arrhenotokous form will be reduced if it produces gregarious broods in which sibmating takes place. Therefore if these 'Allee' effects are important, one would predict that it will be more difficult for thelytoky to go to fixation in highly gregarious species than in species that are solitary. In contrast, it will be relatively easy for a thelytokous form to invade an arrhenotokous population because no conditions will need to be met other than those given in the equation above. This asymmetry in invasion criteria may have important effects on the distribution of thelytokous forms and arrhenotokous forms in populations. Situations may exist in which thelytokous populations can completely displace arrhenotokous populations, even though the number of daughters produced by a mated arrhenotokous female is higher than that produced by an infected female. This may take place when thelytokous populations first arise in marginal areas where the Allee effects make it impossible for arrhenotokous populations to persist. Secondly, these thelytokous populations could then expand into more favourable areas and displace the existing arrhenotokous population by inducing a relative deficit of males. This causes a higher proportion of the uninfected females to remain unmated and leads to a reduction in the average value of x.

The different routes that can lead to a fixed infection in the host population have different effects on the initial composition of the infected population:

1. The population has gone to fixation because the *Wolbachia* that initially infected it caused the infected females to produce a much larger number of infected daughters than the uninfected population produced, i.e. $w\alpha >> x$. The completely infected population will then consist of a number of genetically different clones. These are the result of the transitional phase in which the infected females still mated with the not-infected males. In addition, the females from such recently fixed populations, if exposed to males, should still be willing to mate and fertilize their eggs.
2. The population has gone to fixation through the route which included a mating suppressor. The expectation is that the population consists of much fewer clones. The initially predominant clone should be the one that was the initial mutant. Females from such a population should not be willing to mate any more if they are exposed to males.

Both situations will, in the long run, become similar locally since the 'best' clone outcompetes the rest, assuming a homogeneous environment. However, on a geographical scale genetically different clones could persist. The difference between the situations in willingness to mate should also disappear. Once a population has gone to fixation for infection and the transmission frequency (α) is high, little or no sexual reproduction takes place. All genes involved with sexual reproduction in females and male-specific traits will accumulate mutations. This may be the reason for the lack of mating between males created by antibiotic treatments and infected females from populations fixed for *Wolbachia* infection. Generally, it appears that female function is lost first. In a number of cases males still appear willing to mate, but are not capable of inducing the proper behaviours in females (van den Assem and Povel 1973; Gordh and Lacey 1976; Pijls *et al.* 1996; Silva and Stouthamer 1996). In *Apoanagyrus diversicornis*, this appears particularly clear; females of the thelytokous form are unwilling to mate with males of either the arrhenotokous or thelytokous form, while females of the arrhenotokous form mate successfully with males of either form (Pijls *et al.* 1996). Why genes for female traits may accumulate mutations more rapidly is not clear. Possible explanations are:

1. There are more genes involved in female mating behaviour than in male mating behaviour, and therefore the chance of a female having a mutation on one of those genes is larger. Females would have more genes for mating behaviour than males because mating with a wrong male is more costly for a female than mating with a wrong female is for a male.
2. There may be active selection against expressed traits involved with sexual reproduction once the population has gone to fixation. Some traits involved with sexual reproduction are expected to be costly, for instance pheromone production by females (Pijls *et al.* 1996).

4.10 *Wolbachia* host interactions

In the discussion on the dynamics of the infection I have assumed that the two participants of the interaction might only coevolve through the evolution of nuclear suppressors against the *Wolbachia*. Changes may also take place in the *Wolbachia* genome to enhance its spread through the population.

4.10.1 Increasing the rate of vertical transmission (α)

In most individuals studied in the laboratory, the vertical transmission of *Wolbachia* in *Trichogramma* wasps is high over the first few days of their offspring production. Often later in life more male progeny are produced. However, we do not know how often these effects happen in the field. A *Wolbachia* mutant that causes an increased transmission (α) in its host should spread relative to forms that do not have such an effect. Even in some of the populations where the infection has gone to fixation, the transmission of *Wolbachia* appears not to be complete, at least when these rates are measured in the laboratory. For instance Legner (1985*b*, 1988) found substantial male production in lines of the thelytokous *M. uniraptor* in the laboratory.

4.10.2 Increasing the relative offspring production of the infected females (w)

Wolbachia variants that result in a higher w should spread through the population of infected females. Situations can be envisioned where the *Wolbachia* provides some essential amino acid or some other nutrient to the host. Such a situation seems to occur in *Nasonia* infected with the incompatibility *Wolbachia* (Stolk and Stouthamer 1995). However, we also know that in the case of *Wolbachia* infecting *Trichogramma*, there appears to be a definite fitness disadvantage of the infection.

Another possible scenario for *Wolbachia* to increase w is to manipulate the host into producing fitter offspring. A possibility would be if a positive heterosis exists in a host population for the *Wolbachia* to manipulate the host into fertilizing all of her eggs. Such host manipulation would only benefit *Wolbachia* if the female produced from an unfertilized egg (i.e. a completely homozygous infected female) would have a lower fitness than an infected female produced from a fertilized egg.

4.10.3 Lowering the sex ratio of the uninfected individuals (x)

If infected females are more attractive to males than females that are not infected with *Wolbachia*, this should result in a relatively higher proportion of uninfecteds remaining virgins and therefore producing a lower sex ratio. In addition, if the parthenogenesis *Wolbachia* also causes some incompatibility effects, one could envision that the males sometimes produced by infected females would be incompatible with uninfected females of the population. Therefore under these circumstances even high male production could lead to the spread of the *Wolbachia* population by incompatibility effects.

No evidence exists as yet that the parthenogenesis *Wolbachia* modifies any other behaviour outside that of the chromosomes of infected haploid eggs and that, in some cases, their presence has a negative effect on the offspring production of infected females. Wade and Chang (1995) found that the incompatibility *Wolbachia* in *Tribolium* also enhanced the competitive ability of sperm from infected males, suggesting that additional host manipulation in parthenogenesis *Wolbachia* may also be possible (but see Chapter 2).

4.11 Conclusion

After the discovery, approximately 10 years ago, of the involvement of *Wolbachia* in causing parthenogenesis, our knowledge of this interaction has grown extensively. However, it is also obvious that many questions remain to be addressed. One of the main tasks that remains to be undertaken is the transfer of parthenogenesis *Wolbachia* from infected to uninfected individuals. Once this can be accomplished it will be possible to determine whether the effect of parthenogenesis and incompatibility *Wolbachia* is caused by a host-specific reaction to the presence of *Wolbachia* or whether it is a *Wolbachia*-specific effect. In the former case, in reaction to the same *Wolbachia* some hosts will show the incompatibility effect, while other hosts will exhibit the parthenogenesis effect.

Another area that needs further exploration is the location of the genes coding for the effects we observe in the hosts, i.e. incompatibility v. parthenogenesis. If these traits are encoded on the chromosome of the *Wolbachia*, then how can we explain the disjunct phylogeny of the parthenogenesis and incompatibility *Wolbachia*? At this point it seems that either bacterial chromosomal DNA exchange needs to take place, or the traits are encoded on bacterial extrachromosomal elements, or the step from incompatibility to parthenogenesis evolves easily.

At the cellular level, modification of mitotic division by *Wolbachia* also remains to be studied. The modifications, both in the case of incompatibility and parthenogenesis, appear to be so general that they must take place in some very basic and general process of the first mitotic divisions in the newly laid eggs.

Cytogenetically, it remains unanswered how gynandromorphs come about. A simple shift of the aborted anaphase to one of the nuclei of a later mitotic division seems to be an inadequate explanation for the occurrence of gynandromorphs with very little male tissues.

The only effects of the *Wolbachia* infection that have been detected in forms originating from mixed populations are:

(1) allowing unfertilized eggs to develop into females; and

(2) a negative effect on the potential production of offspring.

Other effects on the behaviour of the infected females have not yet been studied. One possibility is that *Wolbachia* causes the infected female to fertilize all her eggs if there is an heterozygote advantage. Another possibility is that the rare males produced by infected females show cytoplasmic incompatibility with uninfected females.

At the population level the coexistence of infected and uninfected forms requires additional study. How can these two forms persist in a population over time? Two possible explanations could be: frequent horizontal transmission or the existence of suppressor genes in the host population. Horizontal transmission is thought to be rare at best in the field, and no confirmed suppressor genes have been found. The interaction between *Wolbachia* and its host when the population has gone to fixation is also an underexplored field. This should be particularly interesting in some of the

Wolbachia–Trichogramma combinations where fixation should have happened relatively recently, such as *T. cordubensis*.

Finally, parthenogenesis caused by *Wolbachia* has only been detected in Hymenoptera. Is this effect of the *Wolbachia* indeed limited to this insect order or are there other arthropods where *Wolbachia* causes such an effect? Other haplodiploid insects seem likely groups in which to look for parthenogenesis *Wolbachia*.

5 Cytoplasmic sex-ratio distorters

Gregory D. D. Hurst, Laurence D. Hurst, and Michael E. N. Majerus

5.1 Introduction

As previously discussed in this book, cytoplasmic factors in animals, be they organellar, bacterial, or otherwise, tend to be inherited exclusively from the mother. One consequence of this is that cytoplasmic factors will typically have zero fitness when in a male. Hence, a cytoplasmic factor that can either force an increase in the proportion of females in its 'host's' brood, or can otherwise increase the survival and reproductive success of infected female hosts, will spread. To put it another way, Fisher's (1930) prediction of a stable 1 : 1 sex ratio is an analysis of the evolutionary interests of autosomes which does not hold either for sex chromosomes (Hamilton 1967) or for cytoplasmic elements (Hamilton 1979; Cosmides and Tooby 1981). Cytoplasmic genes will be selected to produce a female-bias to the sex ratio. Examples of cytoplasmic sex-ratio distorters have already been discussed in this book, most notably the agents that induce parthenogenesis in a variety of wasps (Chapter 4) and those that are responsible for feminization in crustaceans (Chapter 3).

This chapter examines the remaining incidences. For the most part this means vertically transmitted agents that kill males. However, we start with brief consideration of the (limited) evidence for the existence of other classes of cytoplasmic sex-ratio distorters. Discussion of the population genetics and biology of all of these systems is followed by a body of speculation as to the potential impact of these factors on host evolution.

Prior to this, however, we simply wish to caution that there exist numerous examples of sex-ratio distorters that have nothing to do with cytoplasmic factors. Most notable of these are sex-linked meiotic drive genes. Trivially then, the finding of a female-biased sex ratio in a brood is not proof of the presence of a cytoplasmic distorter. Indeed, neither is the finding of the death of half the progeny, that half all being male (as with male-killers). As Noor and Coyne (1995) have recently found in *Drosophila simulans*, the expression of X-linked genes in the father may kill male progeny (i.e. progeny which do not inherit the X-linked genes paternally). These genes are the sex-linked equivalent of Medea, the maternal effect embryonic lethality trait of the flour beetle *Tribolium castaneum* (Beeman *et al.* 1992). Such genes spread deterministically through a population, particularly if the death of non-bearers increases the probability of survival of their siblings (Wade and Beeman 1994). The demonstration of a cytoplasmic effect thus requires not just the existence of a sex-ratio bias consistent with a cytoplasmic effect; rather it also requires us to exclude other explanations by lineage analysis of the trait.

5.2 Factors biasing the primary sex ratio toward females

Cytoplasmic factors that bias the sex ratio of their host individuals towards the production of daughters spread deterministically through a population. Thus, we see inherited symbionts which act as feminizing factors, and those which act to induce parthenogenesis. There are a number of other mechanisms by which cytoplasmic genes could create such a bias, with a variety of different levels of evidence for their existence.

5.2.1 Increased fertilization frequency in an arrhenotokous species

The best-described example of a cytoplasmic factor that manipulates sex ratios and that is not a male-killer, parthenogenesis-inducer, or feminizer, is a factor called maternal sex ratio (msr) in the hymenopteran *Nasonia vitripennis*. *Msr* is a maternally heritable factor which increases the rate of fertilization of eggs. Given that *N. vitripennis* is haplodiploid, this results in broods consisting of 97 per cent daughters (Skinner 1982). The causal agent of msr has yet to be identified and *msr* is the least well described of all the sex-ratio agents of *N. vitripennis*. Searches for an association of the trait with bacterial presence have all proved negative (Stouthamer, personal communication) and the trait is thus considered to be a product either of a virus, or to be associated with particular mitochondrial types. Its synergism with the male-biasing factor, paternal sex ratio, has been investigated both in theory (Werren and Beukeboom 1993) and practice (Beukeboom and Werren 1992).

A comparable manipulation of reproduction has been described in the bumble bee *Bombus terrestris* (Shykoff and Schmid-Hempel 1991). This species is infected with a maternally inherited trypanosome, *Crithidia bombi*, which affects both queen and worker reproduction. However, its effect upon worker reproduction is more pronounced than on the queen. This asymmetry results in a skew in the sex ratio of sexual progeny produced by the colony. Workers, being unmated, produce sons only (drones). Thus the curtailment of worker reproduction produces a female bias to the sex ratio produced by the colony, which is adaptive for a maternally inherited microorganism.

Agents comparable to these are only possible in organisms where sex is determined by the presence or absence of fertilization (i.e. typically arrhenotokous haplodiploids). However, while monogeny has been found in some such species, reports are otherwise scant. It is possible that the maternally inherited female-biased sex-ratio trait of the haplodiploid mites *Leptotrombidium fletcheri* and *L. arenicola* is analogous (Roberts *et al.* 1977), but the sexuality of the affected females and the nature of the sex-ratio bias (primary or secondary) has not yet been ascertained. The sex-ratio trait appears to be associated with a bacterium, *Rickettsia tsutsugamushi*, in these cases.

5.2.2 Prevention, by cytoplasmic factors in the egg, of Y (or O)-bearing sperm from entering that egg

Assuming a species to be male heterogametic, a cytoplasmic factor could potentially guarantee its transmission to all future generations by only allowing X-bearing sperm to fertilize the egg. Perhaps unsurprisingly, this has never been observed. It has, however, been conjectured as the cause of female-biased sex ratios in the curculionid *Sitophilus oryzae* (Holloway 1985). In an analysis of the sex ratio produced by crosses involving members of different populations of this weevil, the best predictor of sex ratio was a model evoking the interaction of a non-chromosomal maternally inherited factor and a Y-linked gene. There was no correlation between mature brood size and sex ratio, suggesting that male mortality was not the cause. Holloway suggests that this effect may be due to a cytoplasmic gene in the egg preferentially accepting X-bearing sperm. However, he notes that other explanations exist, and the evidence for a cytoplasmic sex-ratio distorter in this case is very far from conclusive.

5.2.3 Prevention, by cytoplasmic factors in the egg, of Z segregating into the egg

If, in a species that is female heterogametic, a cytoplasmic factor influenced meiosis such that the female determining W (or O) chromosome entered a greater proportion of ova, then this cytoplasmic agent would spread. The trait would have the same pattern of inheritance as W (or O) chromosome with meiotic drive, and would be indistinguishable from this by classical linkage analysis.

Doncaster (1913, 1914) investigated what might be an example of this form of manipulation in the magpie moth, *Abraxas grossulariata*. Males and females in the species typically have 56 chromosomes ($n = 28$). Doncaster discovered a strain of all-female-producing females. About half of the females in the all-female broods themselves produced all-female broods, although occasional male production was observed. The females producing the bias appeared to have 55 chromosomes, indicating that they might be XO, rather than the usual XY female. Females with 55 chromosomes producing unbiased broods were also reported, suggesting that this effect was not directly caused by the lack of a chromosome. When examining the female meiosis of normal and female-biased lineages, Doncaster initially reported that in female-biased lineages the inner meiotic plate (that which will form the egg) typically had 27 chromosomes, with the pre-polar body having 28, although in no case were all eggs from a brood shown to have this arrangement. In contrast, females from unbiased lineages with 55 chromosomes displayed a roughly equal division of 27 and 28 chromosomes on the inner and outer plates.

The indication was thus that the sex-ratio bias might occur during meiosis by preferential alignment on the meiotic plate. However, in an incomplete paper written shortly before his death, Doncaster (1922) indicated that this finding was not supported by further analysis. Instead, he argued that the extra chromosome may have been left on the plate during chromosomal separation in female-biased lineages, leaving all the eggs with 27 chromosomes, regardless of the original position of the extra chromosome.

Two possible causes of this sex-ratio bias, parthenogenesis and male death, can be ruled out. With respect to the former possibility, inheritance of paternally derived genes was found. With respect to the latter, analysis of the pedigrees presented by Doncaster (1914) demonstrates that the mean number of females per brood in heavily female-biased broods (including all-female broods) is approximately twice the mean number of females in non-female-biased broods (L. D. Hurst 1993).

One possible explanation that remains is that this represents a case of meiotic drive. Crew (1927) has suggested that the females might have been ZO in constitution. If it really is the case that these females were ZO, then cytoplasmic versus Z drive is also a viable explanation. The case deserves re-evaluation.

The production of all-female broods has been reported for a number of other Lepidoptera (*Acraea encedon*, Owen 1970; *Philudoria potatoria*, Majerus 1981; possibly *Maniola jurtina*, Scali and Masetti 1973). These have been cited as possible incidences of W meiotic drive (Hurst and Pomiankowski 1991). However, in Lepidoptera the W and the cytoplasm behave more or less as a single linkage group and thus any claimed W drive could equally well be a cytoplasmic effect. Thus, in the case of *Acraea encedon*, Owen and Smith (1991) have argued that although W drive is the most likely cause of the female bias, a cytoplasmic effect cannot be ruled out. In both the case of *Acraea encedon* (Owen and Smith 1991) and *Philudoria potatoria* (Majerus 1981) the mortality pattern of individuals in all-female broods did not differ from those in broods with normal sex ratio. This observation rules out male death as an explanation for the trait in these cases. However, cytoplasmically induced meiotic drive remains a tenable hypothesis. For discussion of other potential examples, see L. D. Hurst (1993).

5.3 Factors biasing the primary sex ratio toward males

A number of insect species harbour vertically transmitted bacterial symbionts (*Wolbachia*), which cause cytoplasmic incompatibility, the failure of reproduction of crosses between certain types of individuals (Chapter 2). In the majority of instances of cytoplasmic incompatibility the sex ratio is unaffected. However, in those organisms which can produce males by parthenogenetic means, but require fertilization to produce females, the result of mating an infected male with an uninfected female (or a female infected with a different strain of the bacterium) is the production of all-male broods. In *Nasonia vitripennis* this distortion is not due to the death of females. Rather, the microorganism induces the condensation of the paternal genome in an uninfected background, leaving an embryo that would have been a sexually produced diploid (female) to develop as an asexually produced haploid (male) (Ryan and Saul 1968; Richardson *et al.* 1987; Breeuwer and Werren 1990). The alteration of sex caused by *Wolbachia* is of no adaptive significance to the perpetrator of the bias when considered as an alternative to the strategy of killing the female. Rather, the extra production of uninfected males inhibits the spread of the causal agent.

Cytoplasmic incompatibility is also observed in the arrhenotokous mite *Tetranychus quercivorus*. Here, rather than females being converted into males, they die in embryogenesis. In this species, death occurs because chromosome condensation is incomplete (Gotoh *et al.* 1995*a*). Sexually produced embryos are haploid for some chromosomes, but diploid for others. The involvement of microorganisms in this trait has not yet been proven. Antibiotic treatment was largely performed on females, where there was (unsurprisingly) no restoration of compatibility following treatment (Gotoh *et al.* 1995*b*). Males were treated with a more restricted range of antibiotics. Although there is some evidence for curing from this data, the evidence does not amount to proof.

There is also one example of an inherited virus causing the production of a male-biased sex ratio. This virus, which infects between 0.9 and 5.4 per cent of *Glossina pallidipes* individuals (depending on population) (Otieno *et al.* 1980), causes salivary gland hyperplasia, and appears to interfere with the process or products of meiosis. Males of *Glossina pallidipes* infected with the virus are usually sterile. However, a few have functional testes. If these infected males are mated to uninfected females then a male-biased sex ratio (71 : 29) results

(Jaenson 1986). Infected females mated with uninfected males produce even sex ratios, as do uninfected pairs. Female longevity is reduced by the virus but the females are not sterile and the virus is vertically transmitted through the female lineage with about 95 per cent efficiency. A slight degree of paternal inheritance of the virus is also implicated. Sperm count is low in infected males, implying that the mechanism of action might involve the depletion of X sperm over Y sperm. An adaptive account for this sex-ratio bias is again probably inappropriate.

5.4 Male-killing

There are many records of cytoplasmic symbionts which show host sex biases in their mortality effects, specifically causing heavier mortality on males which inherit them. These 'male-killers' fall into two categories. The first class is the so called 'late-killers'—so called because male mortality occurs late on in development (typically late larval instars). Here, the death of the male is associated with the rupturing of the cuticle during the release of infective spores. Male-killers in this class may more legitimately be regarded as conventional parasites. Why it is legitimate to include them here is that they are less virulent in females, and can be vertically transmitted.

The second category are the 'early-killers' (so called because male mortality is typically during embryogenesis or the first larval instar). Like late-killers, these factors are also maternally inherited. However, the *act* of early male-killing is not usually associated with the horizontal transmission of the agent. For the most part these simply kill males and, in the process, cause their own mortality. Where horizontal transmission is found, it is not associated with the killing of the males. The population genetics of this class of male-killer is different from that of late male-killers. The latter both receive and depend upon the substantial levels of horizontal transmission they receive following the death of males.

5.4.1 Late male-killing

Late male-killers have so far only been recorded in mosquitoes. Many (possibly all) populations of mosquitoes are infected with vertically transmitted microsporidian parasites that can cause lethal infections in fourth-instar larvae. Infection frequencies in the wild are typically anywhere from 2 to 40 per cent. In some species only the males develop lethal infections, and the females survive to vertically transmit the parasite. In others, however, both males and females die, and sometimes, more enigmatically, neither.

In both of the latter two cases, the agent is not a sex-ratio distorter and so we shall not deal with these (reviewed by L. D. Hurst 1993), except to note that whereas a given species of microsporidia was thought either to induce female lethality (i.e. kill all hosts) or be benign to females (i.e. be male-killers), this is not always the case. In the mosquito *Aedes cantator*, for instance, the action of the microsporidian in females is variable, sometimes killing them, at other times developing more benignly and transmitting vertically to the next generation (Andreadis 1991).

Horizontal transmission of these microsporidia is achieved by one of two routes. Either the spores released by dead males infect a copepod intermediary, or they infect other mosquitoes directly. Typically, any given microsporidian species does one or the other. However, Bechnel and Sweeney (1990) present evidence suggesting that *Amblyospora trinus* might be transmissible by both routes into *Culex halifaxi*. The same parasite might also be transmitted from mosquito to mosquito by cannibalism (Bechnel and Sweeney 1990). Mosquito larvae acquire the protists by ingestion. The first stages of the infection in the mosquito involve the protists invading the epithelium of the gastric caeca. From there they invade oenocytes. In such newly infected hosts the microsporidians typically develop in a benign manner in both sexes (although see Andreadis 1988) and are vertically transmitted to the next generation through the eggs. It is typically only in this next generation that male-lethal infections arise.

The mechanism by which microsporidians 'know' whether they are in a male or female is uncertain. Analysis of spore development of *Amblyospora* sp. in *Culex salinarius* gynandromorphs was unable to demonstrate whether hormonal or genetic cues are used (Hall 1990). To be vertically transmitted the protists move from oenocyte to ovary after the taking of a blood meal by the adult host. Diploid spores of *Thelohania legeri* in *Anopheles quadrimaculatus* remain in the oenocyte which dislodges and migrates to the ovaries (Hazard and Anthony 1974), whereas the diploid spores of *Amblyospora* sp. in *C. salinarius* enter the haemolymph and migrate to the ovaries (Hall 1985). Either way, transfer to the next generation is achieved.

The study of microsporidia in mosquitoes raises two questions. First, why are males killed late in development? Secondly, why is there such a diversity of patterns of infection? The answers to both of these questions are unclear. The potential for horizontal transmission permits an explanation for the late timing of death (Hurst 1991). After the fourth instar, mosquitoes pupate in the water. After pupation the adult mosquito is only briefly associated with the water. If the pupal case presents a barrier to release (this fact is uncertain), then the fourth instar would be the last stage during which the microsporidians could escape from their host to water and hence to the intermediate host. In this case, the killing of larvae at the fourth stadium would be seen as an optimizing strategy. Given that vertical transmission is not possible in males, the parasites 'wait' until the host is as large as possible before entering the full-blown infectious state, thereby maximizing spore number.

Extension of the same analysis suggests a reason for the variety of host–parasite interactions. Hurst (1991) has argued that the key parameters are probably the relative efficiencies of vertical and horizontal transmission. The protists in males have no 'incentive' not to kill their host. However, if vertical transmission through eggs is poor, as it often is, and adequate copepod vectors are available, then killing females as well as males can be the best strategy. If, however, protists are assured better chances of transfer to the next generation by vertical rather than horizontal transmission then male-lethal infections are to be expected. Hence we might predict that male-killer organisms should be found where vertical transmission is efficient and copepods are not abundant at the appropriate time, and both male and female lethality is expected under the opposite conditions. Sweeney *et al.* (1989) have suggested that the infections that are benign in both sexes might be associated with a scarcity of intermediate hosts, thus forcing vertical transmission, this being the best route to the next generation. However, this does not explain why males are not killed. One possibility is that a degree of transmission through adult males is possible, during copulation. Though not recorded for mosquito microsporidia, microsporidia in other insects have been observed to transmit from male to female during copulation, and thence to progeny (e.g. Kellen and Lindegren 1971; Armstrong 1977). Although sexual transmission is the exception rather than the rule for insect microsporidia, this possibility should be investigated in the mosquito systems.

Although the cases of late male-killing so far described are restricted to mosquitoes, sex-dependent activity of vertically transmitted parasites is not. Some helminths of mammals have specialized so as to be capable of vertical transmission through milk and across the placenta (Shoop 1991). In females the helminths typically remain latent until pregnancy, at which point they reactivate and migrate to the uterus and mammary glands, from whence the host's progeny become infected. In male hosts, however, they undergo somatic migration upon entering the host, and mature to adulthood. Once adult, they shed eggs which can then be horizontally transmitted.

5.4.2 Early male-killing

Early male-killing elements differ from late male-killing elements in the import of horizontal transmission in their population biology. In only one case, the son-killer of *Nasonia vitripennis* (Huger *et al.* 1985; Skinner 1985), is horizontal transmission known. More ambiguous evidence exists for three other systems (discussed by Hurst and Majerus 1993). The lack of horizontal transmission is consistent with the difficulties encountered in culturing the relevant symbionts. Notably, the exception (the male-killer in *N. vitripennis*) is both extracellular, easily culturable, and transmitted by larval feeding, rather than transovarially (Huger *et al.* 1985; Werren *et al.* 1986; Gherna *et al.*

1991). We may fairly safely state that, with this exception, horizontal transmission is of minor import in the population dynamics of these symbionts.

In terms of population biology, therefore, most early male-killers are directly comparable to agents causing cytoplasmic male sterility in hermaphrodite plants. A variety of theoretically adequate models could be evoked to explain the existence (and persistence) of both early male-killers and cytoplasmic male sterility (note also that the same theory may describe part of the reason for the spread of late-killers) (Lewis 1941; Werren 1987; Frank 1989; Hurst 1991).

From a collection of five parameters a variety of minimal models may be envisaged. These five parameters are: the vertical transmission rate, a; a direct effect on infected female survivorship and fecundity, s; a rate of inbreeding, F, and consequential inbreeding depression $(1-t)$; and an enhanced fitness of female progeny that follows as an indirect consequence of male mortality. Following male mortality, female viability is increased by a multiple b (b being a function of the number of dead males, i.e. a function of a). This indirect advantage may either be due to simple reduction in the degree of antagonistic interactions experienced by females (e.g. reduced probability of being cannibalized, reduced levels of direct sib–sib competition for food), or due to increased levels of resources going to the females due to the death of males (e.g. following consumption of the egg bearing the sibling male). Starting from the null condition that $s=0$, $b=1$, $a<1$, $F=0$, $t=0$, then the following modifications allow persistence of the trait.

1. If s and a are allowed to vary, then the frequency of the factor in the next generation (p') as a function of its present frequency is given by:

$$p' = ap(1+s)/\overline{w}$$

where
$$\overline{w} = p(a(1+s)+(1-a))+1-p$$

and invasion is possible if
$$a(1+s) > 1.$$

This leads to a stable equilibrium frequency at $p* = (a(1+s)-1)/as$.

It follows that, if vertical transmission is perfect ($a=1$), and the agent is neither directly deleterious nor advantageous ($s=0$), it is perfectly neutral and hence may spread by drift. Alternatively, if the agent is directly advantageous ($s>0$), then deterministic spread occurs so long as the rate of vertical transmission (a) is not too low. Under both of these circumstances, the activity of male-killing is not of any relevance as regards the population genetics of the trait. Rather, it is an incidental side-product. Although not aesthetically attractive as a hypothesis (one is left wondering how the factor came to evolve sex-specific activity if this was irrelevant), it is theoretically possible.

Invasion of the factor will result in stable fixation if $a=1$. Naturally this also means that the population will typically go extinct if all infected males die. Otherwise, so long as $s<1$, an internal equilibrium is found.

2. Spread is possible in the absence of any direct fitness benefits ($s \leqslant 0$). In these cases (2a and 2b), the spread of the cytoplasmic factor may be considered an incidence of kin selection (Hurst 1991) and the death of the males is vital to the spread of the trait. Spread will occur:

(a) In the absence of inbreeding effects, if b is adequately high to compensate both for reduced female viability due to direct costs ($s < 0$) and the progressive loss of the factor due to less than perfect vertical transmission ($a < 1.0$), then under these conditions:

$$p' = abp\,(1+s)/\overline{w}$$

where
$$\overline{w} = pb\,(a\,(1+s) + (1-a)) + 1 - p$$

and invasion is possible if
$$b > 1/(a\,(1+s)) > 1.$$

Leading to a stable equilibrium frequency at

$$p* = (ab\,(1+s) - 1)/(b\,(1+as) - 1).$$

This again is unity if $a = 1$, otherwise it reduces to an internal equilibrium.

In addition, and more realistically, as b is a function of the rate of vertical transmission (the higher the rate, the more males dying, so the greater the benefit to survivors), the b term can be replaced by some function of a and parameters to describe the effect of the proportion of progeny dying on the lifetime reproductive success of remaining individuals. This may reasonably be supposed to follow a sigmoidal function obeying the law of diminishing returns (Freeland and McCabe 1997). Under these circumstances the equilibrium frequency can be highly sensitive to small alterations in vertical transmission rate (Freeland and McCabe 1997).

(b) If there is both an adequately high rate of inbreeding ($F > 0$) and an adequately high inbreeding depression ($t > 0$), then infected females may have a higher than average lifetime reproductive success than uninfected ones, and invasion of the factor may be possible (Werren 1987). Invasion under these circumstances requires also that the probability of mating is not severely affected by male-killing. Most obviously, obligately inbred organisms do not provide the conditions for the spread of male-killing agents. In more general terms, the female hosts must still have a substantial chance of successful mating despite the death of their brothers for an advantage to be gained from inbreeding avoidance.

The conclusion from the above is that male-killers are likely to be present in host species that inbreed regularly, but still lose fitness when they do so. There is a subtle balance here between the rate of inbreeding and the fitness loss that occurs following inbreeding. The expected genetic load varies as a function of the rate of inbreeding. If regularly inbred, the equilibrium level of deleterious recessives is likely to be low, and thus inbreeding depression less profound. However, as Werren (1987) notes, inbreeding depression need not be high to allow invasion by a male-killer when inbreeding is common. Thus, we may fairly safely state that populations where there is an appreciable (greater than 5 per cent) level of inbreeding will allow the spread of a male-killer, so long as infected individuals in such populations do not suffer profound losses from failure to find a mate.

5.5 Is male-killing adaptive?

Some 19 cases of early male-killing have been reported from five insect orders (Table 5.1). These include two cases in one species, the ladybird *Adalia bipunctata*, in which different male-killers are known in different populations (Hurst *et al.* 1992, 1997*a*; Werren *et al.* 1994; Zakharov *et al.* 1996). From studies of these cases, we can attempt to answer one central question: is male-killing an adaptive behaviour for the inherited bacteria which display the trait?

It might be considered that the null hypothesis in these cases should be that male-killing is a neutral by-product. There are two tests of this hypothesis. First, we may enquire whether reductions in frequency arising from inefficient transmission are overcome through direct increases in the fitness of females arising from infection (and horizontal transmission, where this occurs). If this is the not the case, we can infer a benefit to male-killing. The second test is more direct. We may enquire whether there is any direct evidence for an effect of male-killing behaviour on the survival and reproductive success of the female hosts of the symbiont.

5.5.1 Indirect evidence for an advantage to male-killing

The case for an advantage to the symbiont of killing male embryos is strengthened if we cannot account for the dynamics of the symbiont without recourse to an advantage to male-killing. In other words, male-killing is likely to be beneficial to the symbiont if we cannot explain losses through inefficient transmission by increases in the fitness of female hosts in the absence of male-killing.

5.5.1.1 Transmission efficiency

Laboratory studies suggest that male-killing symbionts generally have imperfect transmission. The one exception to the rule of inefficient transmission is the butterfly *Hypolimnas bolina*, in which Clarke *et al.* (1975) reported

Table 5.1 Host factors bearing upon the dynamics of male-killing bacteria

Species	Inbreeding occurrence	Inbreeding depression	Sibling egg consumption?	Antagonistic interactions between siblings?		
				Competition for food	Sibling cannibalism	Other (noted)
Oncopeltus fasciatus	Unknown	Unknown	Yes	Unlikely	No	
Drosophila willistoni	Rare, if any	High	No	Unknown	No	
Drosophila bifasciata	Unknown	Some	No	Unknown	No	
Nasonia vitripennis	Common	Low	No	Probable	Unknown	
Caraphractus cinctus	Common	Unknown	No	Probable	No	
Hippodamia 15-signata	Unknown	Unknown	Yes	Unlikely	Yes	
Harmonia axyridis	Unknown	High	Yes	Unlikely	Yes	
Menochilius 6-maculatus	Unknown	Unknown	Yes	Unlikely	Yes	
Coleomegilla maculata	Unknown	Unknown	Yes	Unlikely	Yes	
Adalia bipunctata/Rickettsia	Rare, if any	High	Yes	Unlikely	Yes	
Adalia bipunctata/Spiroplasma	Rare, if any	High	Yes	Unlikely	Yes	
Coccinella 7-punctata	Rare, if any	Unknown	Yes	Unlikely	Yes	
Ips latidens	Unlikely	Unknown	Probable	Probable	Yes	
Gastrolina depressa	Unknown	Unknown	Unknown	Unknown	Unknown	
Epiphyas postvittana	? Common	Unknown	No	Unlikely	No	Competition for spin-up sites
Cadra cautella	No	Unknown	Probable	Unknown	Yes	
Hyplolimnas bolina	Unknown	Unknown	Unknown	Unknown	Unknown	
Spodoptera littoralis	Low	Unknown	No	Probably	No	
Estigmene acrea	Low	Unknown	Yes	Possibly	Possibly	

100 per cent vertical transmission efficiency on limited data. In this species, available data do not rule out the possibility that all-female broods result from a Y-linked maternal effect gene rather than an endosymbiont (L. D. Hurst 1993). In other cases, laboratory-measured vertical transmission efficiencies vary from around 80 per cent in the moth *Spodoptera littoralis* (Brimacombe 1980) to over 99.9 per cent (1 in 1014) in *Harmonia axyridis* from Japan (Majerus *et al.* 1997). In one species, *Adalia bipunctata*, in which two male-killers have been found in different populations, the transmission efficiencies of the two are substantially different, that of the rickettsia-like male-killer from western Europe being about 87 per cent (Hurst *et al.* 1993), while that of the spiroplasma from Russian populations is over 98 per cent (Hurst *et al.* 1997*a*).

All of these laboratory measures of transmission efficiency should be treated with caution. Laboratory rearing is often continuous (no over-wintering), with females being well fed and reproducing early, and in environments of regulated temperature. All of these may affect transmission efficiency, and the observation that transmission efficiency declines with time in laboratory culture (see Brimacombe 1980 for example), and that certain laboratory-measured transmission efficiencies are incompatible with the observed frequencies of male-killer infections certainly suggests laboratory measures of this parameter may be somewhat limited in value. Perhaps the most important parameter is temperature. The action and transmission of inherited symbionts are frequently temperature-sensitive and male-killers are no exception (Magni 1954; Malogolowkin 1959). Rearing temperatures in the laboratory are frequently lower than those found in the wild, and laboratory measures of transmission efficiency may be artificially high because of this.

5.5.1.2 Effects on female survivorship and reproduction in the absence of male-killing

Inefficient transmission must be balanced by beneficial effects of infection or horizontal transmission for the act of male-killing to be non-adaptive. We have previously noted that there is little evidence for frequent horizontal transmission, save the case of *N. vitripennis* (Skinner 1985). But what evidence is there with regard to the direct effect of infection? A negative effect of infection with a male-killer on female survival and fecundity has either been found or (more anecdotally) noted in five of seven cases studied. In the moth, *Epiphyas postvittana*, profoundly reduced egg production was noted in laboratory stocks (Geier *et al.* 1978). In another moth, *Spodoptera littoralis*, field-collected clutches bearing a male-killer were observed to be significantly smaller than uninfected ones, and a direct effect of infection concluded (Brimacombe 1980). Further, Matsuka *et al.* (1975) noted that in their laboratory, infected females of the ladybird *H. axyridis* had higher levels of sterility and lower oviposition rates than uninfected females.

Experimental analysis of direct effects of infection requires control for the rearing density of larvae from male-killed and normal clutches. Such analyses have been performed on four species. In two species, *Drosophila bifasciata* and *Adalia bipunctata* (rickettsia-like male-killer) there was evidence of a negative effect of infection on adult performance (Ikeda 1970; Hurst *et al.* 1994). However, in the neotropical drosophilids, *D. willistoni* and *D. nebulosa*, infection appears to speed larval development (Malogolowkin-Cohen and Rodriguez-Pereira 1975; Ebbert 1991). In spite of this, the question of an advantage is not clear. In *D. willistoni*, the speeding of larval development is accompanied by an increase in adult female sterility and decreases in female longevity (Ebbert 1991). It is also notable that early studies on this system revealed that although male embryos had much higher mortality than females, some female embryos did also succumb to the symbiont; further to this, examination of female embryos from male-killed broods revealed them to have necrotic cells, interpreted as sub-lethal effects of infection (Counce and Poulson 1962). Furthermore, infections naturally decline in laboratory cage experiments (Ebbert 1995). These experiments, taken as a whole, suggest a negative effect of infection, but the extrapolation from laboratory study to field dynamics should be undertaken with great care (for comments on this, see Hurst *et al.* 1994; Ebbert 1995).

5.5.2 Direct evidence for beneficial effects of male-killing

For a male-killer to spread when both vertical transmission is less than perfect ($a < 1$) and females suffer a direct cost to possessing the symbiont ($s < 0$), female progeny from infected female hosts must gain some advantage from the act of male-killing. The models outlined above discussed three possible factors: a reduction in inbreeding, a direct increase in the amount of food available, and a reduction in antagonistic interactions.

5.5.2.1 Decreases in the rate of inbreeding

For an advantage to accrue from decreased rates of inbreeding, the species must normally indulge in some level of inbreeding, and exhibit some level of inbreeding depression. Because of the reciprocal relationship between the rate of inbreeding and the severity of inbreeding depression, both factors have to be measured. However, in only one case, *Adalia bipunctata*, has the appropriate field work been undertaken. Here, inbreeding depression is severe in laboratory stocks, being manifest in a substantial increase in late-embryonic mortality (the dead embryos are visible in newly hatched clutches). Assessment of embryonic mortality in a large number of field-collected clutches found little evidence of such death. It was therefore inferred that inbreeding was very rare in the field, and thus unlikely to be important in the dynamics of the male-killer in this species (Hurst *et al.* 1996*b*).

Past studies have attempted to assess the importance of inbreeding avoidance as a factor by inferring the rate of inbreeding from ecological data (Hurst and Majerus 1993). Even using this indirect approach, an appraisal of the likely level of natural inbreeding is only possible in 11 of the 19 hosts of male-killers. In two of these there is a real likelihood that male-killing could reduce deleterious inbreeding effects. The first of these is the moth *Epiphyas postvittana*. In this species, mating occurs before dispersal (Gu and Danthanarayana 1990), and it is probable that, particularly under conditions of low population density, inbreeding is common. The other species in which inbreeding occurs is the parasitoid wasp *Nasonia vitripennis*. Here, males emerging from host puparia frequently mate with their siblings that emerge from the same host. In these cases, it is likely that the avoidance of host inbreeding will be a significant benefit to a symbiont with male-killing behaviour. However, it is hard to ascertain definitively whether the avoidance of inbreeding will yield a substantial enough increase in fitness to maintain a male-killing symbiont on its own. The level of inbreeding depression in these two species is not known, and in the case of the haplodiploid *Nasonia vitripennis* it is likely to be low. Because males are haploid, selection against rare deleterious recessive alleles is strong even in outbred populations of this species. However, there is still some potential for inbreeding depression because some mutations will be sex-limited, and inbreeding depression may result from deleterious recessives which are expressed only in females. It is likely that inbreeding and inbreeding depression are sufficient in magnitude to maintain the male-killing elements in these species, the levels required being quite modest for a symbiont with high vertical transmission efficiency (Werren 1987); however, definitive proof awaits.

5.5.2.2 Direct access to resources

A direct increase in access to resources clearly occurs in some species. The increase here is from the dead males, in species where dead siblings are consumed. Sibling egg consumption is found in the milkweed bug, *Oncopeltus fasciatus*. Further, the widespread existence of sibling egg consumption in the Coccinellidae (ladybirds) may explain the common occurrence of male-killers in this group (Hurst and Majerus 1993; Hurst *et al.* 1996*a*). To date, six male-killer–host interactions have been found in this group.

The importance of this egg meal has been questioned by certain authors, with Ebbert (1995) notably arguing that the existence of non-sibling egg consumption made this factor unimportant, as resources were directed towards both siblings and non-siblings. However, this argument clearly fails when the ecology of cannibalism is appreciated in any depth. There are two sorts of egg consumption in ladybirds; consumption of unhatched clutches by larvae which happen upon them (Fig. 5.1a), and consumption of eggs which fail to hatch by sibling larvae from the same clutch (Fig. 5.1b). The existence of the former, undirected, form of consumption (in this case, cannibalism) has been incorrectly held to deny the strong sibling bias and advantage to infected larvae from the latter source. Non-sibling cannibalism involves the consumption of both infected and uninfected female embryos by infected and uninfected female larvae in direct relation (both for victim and cannibal) to their frequency in the population. There is thus no effect of this form of egg consumption on the frequency of a male-killer. The second form of egg consumption is strongly sibling-biased. Neonate larvae from male-killed clutches will gain extra egg meals during their early life.

The more pertinent question, and the one which has not yet been fully addressed, is whether the extra egg meal received by females from male-killed clutches significantly increases survivorship. There are good reasons to suspect it does. First, arguing by design, sibling egg consumption would not have evolved were there not some benefit to it. We can go beyond such arguments to examine the ecology of these species for evidence of an advantage. Significantly, in these species there is significant mortality of neonates arising from failure to obtain their first meal. This has been observed in *Oncopeltus fasciatus*, where young larvae have to find milkweed seed pods of a suitable age (Ralph 1977), and in two of the coccinellids, *Adalia bipunctata* and *Harmonia axyridis*, in which neonates must find and subdue aphids, but often fail to do so (Banks 1955; Dixon 1970; Wratten 1973, 1976; Osawa 1992). In the coccinellids, there is some empirical evidence that the egg meal obtained by members of male-killed clutches does increase the probability of neonate survival. Working on the general effects of egg consumption on survival of *Harmonia axyridis* neonates (rather than the effect of male-killing *per se*), Osawa (1992) showed a significant positive effect of egg consumption on survival to the second instar under conditions of low aphid density. Working on the effects of male-killing itself on survival probability, G. D. D. Hurst (1993) observed that females in male-killed clutches had increased survival time in the absence of food compared to individuals from normal clutches (Fig. 5.2). Such larvae will therefore have longer to find suitable aphid prey before starvation. Further, the larvae will be larger when they disperse and so are likely to be able to subdue a greater size range of aphid prey.

A further criticism of the egg consumption hypothesis is that the same correlation between male-killer incidence and egg consumption would arise if infectious transmission followed egg consumption. This is particularly important with respect to non-sibling egg consumption. However, studies on *Adalia bipunctata* feeding on large numbers of eggs from male-killed clutches to uninfected larvae failed to produce any reproducible evidence of adults with

Figure 5.1 Egg consumption in the coccinellid beetle *Adalia bipunctata*. (a) 'Non-sibling egg cannibalism', the consumption of eggs by larvae which previously hatched from a different clutch; (b) 'sibling egg cannibalism', the clutch in question bearing a male-killer; the surviving female larvae are consuming the eggs containing slow developing or dead male embryos. Note that non-sibling cannibalism does not imply that the participants are necessarily unrelated, rather it implies that there is no necessity that the cannibal is related to the cannibalized individual, as is the case in sibling cannibalism. Further, note that the label sibling egg cannibalism does not imply that the egg consumed is alive and viable. Inter- and intraclutch egg consumption would be better expressions to represent these events.

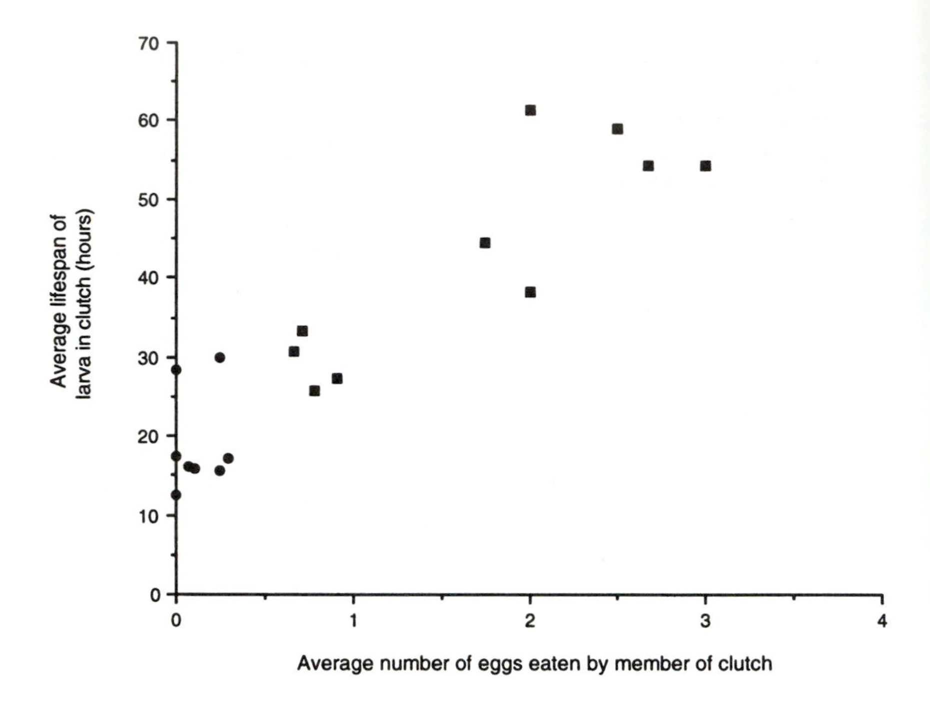

Figure 5.2 The effect of egg consumption on survivorship of neonate larvae in the absence an aphid meal. Male-killed and normal clutches were allowed to hatch and consume unhatched eggs. The number of unhatched eggs consumed, and the number of hatching larvae consuming these eggs was recorded, and the average egg meal obtained by each larva in the clutch calculated. Larvae from these clutches were removed to separate Petri dishes, and their survival time (point until movement ceased) measured. Each point on the graph represents the mean survival time of members of a single clutch, plotted against the average egg meal obtained by members of the clutch. There is a correlation between survival time and the quantity of egg meal on the natal clutch ($r = 0.917$; 16 d.f.; $p < 0.001$). Larvae from uninfected clutches (represented by circles on the graph) live less long than larvae from infected clutches (represented by squares on the graph). Further, larvae from infected clutches where a high proportion of individuals hatched (small egg meal per larva) had a shorter survival time than those where only a low proportion hatched (large egg meal per larva). (Graph reproduced from G. D. D. Hurst (1993) by permission of the author. Full details of methodology may be found therein.)

either cytological evidence of infection or phenotypic evidence of the male-killing trait (G.D.D. Hurst 1993). It thus seems unlikely that infectious transmission following egg consumption explains the frequency of male-killing, at least in this coccinellid.

5.5.2.3 Decreases in the level of antagonistic interactions

The alternative advantage to a symbiont in killing males is the reduction in future antagonistic encounters suffered by females. If there are antagonistic interactions between siblings, then the females arising from male-killed clutches will suffer these less and thus have increased probabilities of survival to reproduction.

Antagonistic interactions between siblings are commonly observed in species bearing male-killers, and this 'benefit' to male-killing is probably widespread (though not universal: see Hurst and Majerus 1993 for a full discussion). Competition between siblings for a food resource is likely to be the most common. In bark beetles, where larvae from a single female are in close proximity, interference is inevitable. In a parasitoid where many eggs are laid in one host, the death of males may increase access to resources for the remaining females (Skinner 1985). In spite of this expectation, Balas *et al.* (1996) found no evidence for size differences between individuals derived from male-killed and normal clutches in *Nasonia vitripennis*.

The avoidance of antagonistic interactions should be seen in a wider context. In the moth *Epiphyas postvittana*, for instance, siblings dispersing from egg clutches compete for limited spin-up sites amongst the leaves of a variety of host plants. The mortality during dispersal is very high (MacLellan 1973; Geier and Briese 1980), and the distance that larvae have to travel to find a safe feeding site is negatively correlated with their survival (Danthanarayana 1983). The lower density of dispersing larvae from clutches laid by females infected with the male-killer would likely give female progeny from such clutches an increased probability of securing a spin-up site.

It need not be assumed that antagonistic interactions necessarily involve competition for a limiting resource. Cannibalistic interactions occur in a wide range of species. In ladybirds, for instance, early-hatching larvae consume late-hatching ones. In male-killed clutches, there are two factors which are likely to reduce cannibalism rates upon late-hatching females. First, there are fewer early-hatching larvae to do the cannibalizing, the males which would have been in this class being dead. Secondly, when cannibalism does begin, a late-hatching female is surrounded by dead males. These may be preyed upon before her (this is perhaps likely by weight of numbers). A dilution effect therefore exists. In the laboratory, at least, females in male-killed clutches have a lower rate of being cannibalized than females from normal clutches (G. D. D. Hurst 1993). These types of cannibalistic interactions are common in ladybirds, and are also found in *Ips latidens* (Miller and Borden 1985), a bark beetle that is host to a male-killer. However, in this species sibling cannibalism is not generally larva upon embryo; rather it is larva upon larva.

The above discussion shows promise for future research. Male-killing does appear explicable as an adaptive trait for the symbiont in the majority of host species. It is unlikely, for instance, that the number of male-killers found in the coccinellid beetles is chance; more likely, it is the habit of consuming dead and late-hatching eggs, found in this group, that promotes it. It is probable that other groups with similar habits, such as the chrysomelid beetles, will show a similar density of infection with male-killers. This said, it would be premature to say that we fully understand, in even one case, the dynamics of male-killing. Careful field work across a range of species is required.

5.6 The causal agents of male-killing

The discussion above concentrates on the dynamics of male-killing strains. The major conclusion from this is that certain factors associated with host ecology are likely to be important in dictating the incidence of male-killing bacteria within the Insecta. We can consider also the reverse side of the symbiosis; namely, are there limits on the evolution of male-killing behaviour by symbionts?

Most recently, molecular phylogenetics has allowed resolution of the affiliation of symbionts (Chapter 1). It appears that early male-killing, in contrast to CI, parthenogenesis-induction, and feminization, is the product of bacteria from across the range of the eubacteria (Table 5.2). Early male-killing has had many independent evolutions.

The diversity of agents which have and can evolve male-killing behaviour suggest that inherited bacteria may generally have the ability to mutate to a strain with male-killing behaviour. This diversity parallels the diversity of bacteria which are beneficial to hosts through playing a role in host metabolism. Inherited bacteria may thus all face alternative paths, the path followed being dependent upon their host. They may be either parasitic

Table 5.2 Current knowledge of the phylogenetic affiliation of male-killing bacteria

Host species	Bacterium	Reference
Drosophila willistoni (Diptera)	Spiroplasma, Gp. II (Class Mollicutes)	Hackett *et al.* (1985)
Nasonia vitripennis (Hymenoptera)	*Arsenophonus nasoniae* (Enterobacteriaceae) (γ-Proteobacteria)	Werren *et al.* (1986) Gherna *et al.* (1991)
Adalia bipunctata (Coleoptera) (Cambridge popn)	*Rickettsia typhi* relative (unnamed) (α-Proteobacteria)	Werren *et al.* (1994)
Adalia bipunctata (Coleoptera) (Russian popns)	Spiroplasma, Gp. VI	Hurst *et al.* (1997*a*)
Harmonia axyridis (Coleoptera)	Spiroplasma, Gp. VI	Hurst *et al.* (1977*a*)
Coleomegilla maculata (Coleoptera)	Sister group to genus Blattabacterium (Class Flavobacteria)	Hurst *et al.* (1996*a*, 1997*b*)

(through male-killing) or beneficial (through performance of an anabolic role). This has interesting evolutionary connotations in the wider context of inherited symbioses in the Insecta. In certain hosts, both parasitic and beneficial symbiont roles may be possible (Chapter 1). Transition from a symbiont playing a host-beneficial metabolic role to one which is a parasitic male-killer (and vice versa) may be a real possibility. However, the conditions of host ecology that promote the evolution of male-killing behaviour may preclude the development of long-term beneficial symbioses.

Evidence for this thesis comes from a recent study of male-killing in the ladybird *Coleomegilla maculata*. The agent of male-killing in this species has been identified as a flavobacterium, and phylogenetically forms the sister group to *Blattabacterium*, the beneficial symbiont of cockroaches (Hurst *et al.* 1997b). The two closely related bacteria have very different relationships with their two host species. This is at least suggestive of the possibility that symbionts may be able to switch from being mutualistic to parasitic with relative ease.

5.7 The mechanism of male-killing

The reason for the diversity of male-killing elements is currently a matter for conjecture. The most appealing hypothesis perhaps is that male-killing is a relatively simple trait to evolve. Male-killing is sex-limited pathogenesis; the key step in the evolution of male-killing is merely how pathogenesis comes to be sex limited. At this moment, it is uncertain how (or even if) the symbionts detect sex. Two hypotheses appear tenable. First, bacterial behaviour may alter in response to a host cue of sex. The bacteria only grows or produces toxin (or does whatever effects male death) in males. Alternatively, bacterial behaviour may be constant, but the effect of bacterial infection is dependent upon sex. For instance, interference with the process of dosage compensation or male sex determination might produce sex-specific death. Here, the bacteria behaves identically in male and female hosts but its effects are limited to the male.

It is known that the cue for male-killing varies between species. In the arrhenotokous *Nasonia vitripennis* fertilization is used as the cue for pathogenesis. This species is host to a parasitic B chromosome termed paternal sex ratio (PSR) which, when inherited from the male, causes the paternal chromosome set to condense during early development (Werren *et al.* 1981; Nur *et al.* 1988; Werren 1991). The result is an individual that has been fertilized, but is still male. These individuals are not killed by the male-killing agent. Given this could not occur in a diplodiploid species, we can say that the cue used by the bacterium differs between host–symbiont interactions. A survey over a wider range of hosts is required to determine how many ways exist.

Little is known of the mechanism of male-killing, but it does appear that male death is often associated with bacterial proliferation. Male death in *Drosophila melanogaster* is associated with the concentration of spiroplasma in the male ventral nervous tissue (Tsuyichiya-Omura *et al.* 1985). These observations tend to suggest that bacterial behaviour is different in the two sexes, but can we rule out constant bacterial behaviour? The alternative explanation is still viable. One can imagine a bacterium producing a toxin in both sexes which has the effect, in males, of influencing host physiology to allow it to proliferate. One molecule would therefore produce different phenotypes in male and female hosts.

5.8 Discussion

Here we want to move away from a simple descriptive analysis to questions of broader and more general interest. We will concentrate on three issues. First, although it does not often seem to be recognized, the cytoplasmic male-killers are an exquisite testing ground for theories of virulence. Secondly, clade selection may be important in determining features of cytoplasmic sex-ratio distorters and can add to the growing literature that sees higher-level selective processes as shaping the extant fauna and flora.

Finally, it is important to ask whether cytoplasmic sex-ratio distorters are 'influential passengers' or merely curios to be relegated to the back page of evolution. In this context we ask whether male-killers might lead to changes in sexual selection, the spread of sexually transmitted diseases, and clutch-size evolution. In addition, we ask whether male-biasing agents could predispose their hosts to the evolution of eusociality, and whether the persistence of male-killers will upset mitochondrial-based molecular phylogenies.

5.8.1 Male-killers and the evolution of virulence

In the recent past it was the accepted wisdom that disease-causing agents should evolve to be of lower virulence. The reasoning was group selective: if parasites were not avirulent they might cause their own extinction by causing the extinction of the host. The correct question to ask, however, concerns the fate of a parasitic mutant that increased its virulence. It was realized that such mutants can spread under numerous conditions and hence there need be no general tendency to reduced virulence.

Following Price (1972) it has often been noted that the key determiner of optimal virulence was the fitness covariance between the parasite and the host: if the fitness of the parasite positively covaries with that of the host, then it is in the 'interest' of the parasite to keep the host alive or at least to minimize harm (Frank 1996). This relationship between cotransmission (or what more ecologically may be thought of as codispersion), fitness covariance, and the evolution of virulence (or avirulence) is implicit, and often explicit, in much of the early consideration of intragenomic conflicts and hierarchical selection (Lewis 1941; Östergren 1945; Hamilton 1967; Price 1970, 1972; Leigh 1971, 1991; Alexander and Borgia 1978; Eberhard 1980; Cosmides and Tooby 1981; Frank 1983; Wade 1985). The same understanding has been largely independently derived by more 'classical' parasitologists.

From this logic it is often supposed that vertically transmitted factors should have zero virulence. This is not so for several reasons. First, and most trivially, vertical transmission efficiency of a maternally transmitted symbiont may be correlated with cost to the host. It is trivially not the case that the symbiont should be selected to have no virulence, as this would equate to no transmission.

More interestingly, as regards this chapter, the positive fitness covariance between host and symbiont is only between *female* host and symbiont. Classical theory of virulence would thus predict that parasites that could escape from male hosts but be avirulent in female ones would be at an advantage. This seems to be what numerous microsporidian parasites of mosquitoes do. The same is true of the helminth infections in mammals, in which the parasite activates dispersal behaviour in males but lays dormant in females.

In the above instances, the harm to males is, as so often assumed in theories of virulence, incidental to the parasite. However, in the case of early male-killers it looks as though male death may well be critical to the spread of the trait. In ladybirds, females eat their dead brothers thereby gaining a first meal. If reduction in inbreeding is relevant to the spread of male-killers then death *per se* is not important. Sterilization of males would have the same effect. If death is important for reasons of reproductive compensation, then male-killers may be expected to kill the males as early as possible.

Although it is important not to take a group-selective approach when theorizing about virulence, it may be the case that higher-level selective processes do determine features of extant systems. Consider, for example, the problem of the evolution of vertical transmission rates of male-killers. Given the proviso that transmission efficiencies are known from laboratory measures only, and hence may be flawed, it is intriguing that cytoplasmic sex-ratio distorters are unusual among symbionts in often not being perfectly vertically transmitted.

5.8.2 Clade selection and vertical transmission?

The 'optimal' male-killer, in terms of rate of spread, would be transmitted to all of the progeny of a female, kill all the sons, and have minimal direct effect on host female survivorship. Given that the most advantageous allele is the one most likely to invade, one might expect that male-killers with maximal spread rates (i.e. closest to this Darwinian daemon) would be more common than those with lower spread rates. However, vertical transmission of male-killers is typically not perfect. It might be argued that imperfection was the result of constraint; the symbiont doing the best it can, in evolutionary terms. However, this argument is undermined by the finding that non-male-killing inherited symbionts (CI-inducing *Wolbachia*, beneficial symbionts) have vertical transmission efficiencies that are typically near-perfect. Constraint is hence not an attractive hypothesis.

One possible explanation for below-perfect vertical transmission is clade selection. Were vertical transmission perfect, then males could easily be lost from populations, at which point the populations are likely to go extinct. There is hence a higher-level filter that directly affects the vertical transmission rates that we presently see.

However, although clade selection is one explanation for the inefficient transmission of male-killers, individual (symbiont)-level explanations may also account for this observation. Inefficient vertical transmission may be an adaptive peak for the symbiont if there is a trade-off between transmission efficiency and virulence (i.e. if high transmission is accompanied by high costs to the female host). Hurst *et al.* (1994) discuss why such a trade-off might exist. If a host germline is not easily accessible to bacteria (and there is little reason to suppose it is designed to be so), then transmission efficiency and bacterial density in the host may be correlated. Further, a correlation between bacterial density and virulence to the host is not unreasonable. Thus follows a trade-off, a situation where inefficient transmission is adaptive to the symbiont.

For this hypothesis to be a good explanation for the low vertical transmission efficiency of male-killers, it must account for the observation that male-killers have lower vertical transmission efficiencies than CI-causing and beneficial symbionts. Male-killing differs from these traits in that male-killers select for modifiers in the host genome that prevent their transmission, whereas in the case of CI and beneficial symbioses, modifiers in the host genome which increase symbiont vertical transmission rates may, at least in theory, invade (Turelli 1994). The lower transmission rates may hence be a consequence of antagonistic coevolution of host and male-killer.

The case for host resistance in the form of selection against transmission of the bacterium requires detailed experimental examination. It is clear from studies of male-killers in drosophilid hosts that there is genetic variation within the host for the probability of transmission of male-killers to progeny (Calcavanti *et al.* 1957; Malogolowkin 1958). What is less clear is whether selection has acted upon this genetic variation, resulting in the observed transmission efficiency of the symbiont. This awaits further study.

However, even if non-perfect vertical transmission rates are the results of host suppressors of vertical transmission, one is still left the problem of why it is that host modifiers of the vertical transmission rate do not cause abolition of vertical transmission and hence lineage purification. As isopod crustaceans have developed the capability to control vertical transmission of the F bacterium, a constraint explanation again seems unreasonable. A cost to modification is a potential explanation. Group selection in favour of a female-bias may also be of some importance. Population structure and the window of opportunity of colonizing individuals has been shown to be of importance to the maintenance of cytoplasmic male sterility in *Thymus vulgaris* in southern France (Mannicacci *et al.* 1996).

5.8.3 The effects of infection upon host evolution

Given the possibility that male-killers may be coevolving with their host, we may now ask what effects male-killers have on host evolution and whether evidence of coevolution exists. First, we may consider the effects of selection to ameliorate or avoid the effect of the parasite. Secondly, we may consider effects of selection that arise from infection, but do not directly alter the frequency of infection. Finally, we may consider unselected consequences of infection, such as the effect a male-killer has upon the evolution of the cytoplasmic genome, with which it is in linkage disequilibrium.

5.8.3.1 Selection against infection

Male-killing symbionts are likely to be parasites (direct benefits appear to be scant, and resource benefits are unlikely to completely compensate for the loss of half of ones progeny). As such, they provide the conditions for the spread of modifiers that suppress their action and (as discussed above) transmission.

Resistance to the action of male-killers (individuals with symbionts where male-killing is ineffective despite transmission of the symbiont) has not been recorded in any system. It is unsure whether there really is a lack of resistance to male-killing in natural populations, or whether the correct studies have yet to be undertaken.

There is one argument that suggests that the lack of observation of resistance to the action of male-killing symbionts represents a real lack of resistance alleles. This is that it may be difficult for resistance to evolve through the gradual accumulation of certain modifiers of small effect. One would imagine that a modifier that somehow delayed death until after male reproduction would be able to spread. But what about a modifier that only slightly delays the timing of male mortality? Would this spread? The curious answer is that it would not and hence the gradual evolution of delay of mortality would not be possible.

To understand why this is so, consider a situation where the advantage to male-killing is in reproductive compensation, through consumption of the eggs containing the dead males. Consider further a situation where the males cannot, initially, prevent their own mortality. If this is the case then selection on the host will favour the early mortality of the male, so maximizing reproductive compensation. In *Adalia bipunctata*, for instance, newly laid eggs represent a more significant nutrient package than eggs in which embryonic development is significantly progressed (Agarwala 1991). This is akin to brood reduction in logic. If being killed by a bacterium is inevitable, it is best for the affected male ladybird to speed its rate of death, and thus maximize the reproductive compensation to sisters. A modifier slightly delaying male death would be associated with reduced kin benefit but no gain in male transmission of the modifier. It would hence not be able to invade.

In addition to this, the fact that a male killed late in life is less nutritionally valuable than one killed early means that some modifiers which save a proportion of males by altering the distribution of male survival time do not spread. Consider a modifier that results in the survival of one male, where this survival is accompanied by a longer survival period of the other males, which still die prior to reproduction. This modifier will often not spread, for the gain of one male is associated with the loss of nutrition to its sisters. The gradual accumulation of modifiers of this kind would also be difficult.

Selection against the effects of infection may follow one further pathway, not previously discussed in the literature. High levels of male-killer infection in a population may promote the evolution of female-biased primary sex ratios or even parthenogenesis. Host females with the bacteria who produce exclusively female progeny never bear the cost of male mortality, even if they are infected. Uninfected females may then be expected to produce male-biased sex ratios to compensate. This pathway relies on a female being able to alter sex allocation in response to a cue of infection. There is little empirical evidence to support this conjecture, save the frequency of symbionts being associated with haplodiploidy and with biased sex ratios (Hamilton 1993). It remains a plausible, but in no sense established, pathway of resistance.

5.8.3.2 Evolutionary effects of male-killer infection and other cytoplasmic sex-ratio distorters

The presence of a male-killer in a population alters the population sex ratio. With high infection rates, the change is quite profound. Shifts in the population sex ratio have several evolutionary consequences. We will discuss three: effects on sexual selection, on dynamics of sexually transmitted diseases, and on clutch size. Finally, we look at the effect male-biasing agents may have on the evolution of eusociality.

Sexual selection

When males are common, selection favours choosy females. Concomitantly, selection favours males which invest time and energy competing with other males for access to females, favours soliciting female attention with gifts, and guarding of paternity by overproduction of sperm, sperm guarding chemicals, or whatever. However, in a population infected with a male-killer, males become more rare. With progressively more female-biased sex ratios, selection moves away from a situation where females are choosy and mate soliciting, and guarding strategies are optimal to males, to one where female choosiness declines and less time is spent by males on soliciting females and assuring paternity, to one where male choosiness evolves.

It is thus clear that male-killers, by influencing population sex ratio, may be an important factor in the evolution of the hosts mating behaviour. Beyond this, these changes in mating behaviour may be important in the dynamics of the symbiont. Consider the spread of a male-killer with high vertical transmission efficiency. In this case, population extinction is a real possibility. However, at high infection rates (very female-biased population sex ratio), male choosiness may evolve. The possibility then exists for selection to act to favour males that choose to mate with uninfected females. A stable equilibrium level of infection is possible: decreases in infection rate produce less female-biased sex ratios, selection for lowered male choosiness follows, leading to increases in infection rate; increases in the proportion of females infected produces more female-biased population sex ratios, selection for increases in choosiness by males follows, leading to decreased infection rates. This selection for male choosiness at high infection rates could therefore prevent the extinction of a population. However, the crucial element (the ability of males to discriminate between females on the basis of infection) remains a matter of conjecture. In addition, it is unsure whether the rate of spread of male choosiness would be great enough to prevent population extinction.

Sexually transmitted diseases

Sex ratio is an important parameter in the dynamics of sexually transmitted disease (STD); slightly biased sex ratios produce a situation where the average number of mates per male increases without substantially altering the mean number of partners engaged by a female. This would increase the incidence of a sexually transmitted disease in the population. Clearly, very female-biased sex ratios, where the mean number of partners engaged by a female decreases, produces the opposite effect. These effects are certainly not globally important, but are a reality in the ladybird, *Adalia bipunctata*, which possesses both a male-killing bacterium at low frequency and an STD (Hurst *et al.* 1995).

We can create similar arguments for the spread of a segregation distorter; the spread and equilibrium levels of a driving chromosome depend upon the intensity of inter-male sperm competition; high levels of promiscuity produce intense inter-individual sperm competition, and may prevent or slow the spread of certain meiotic drivers (Haig and Bergstrom 1995). A slightly female-biased sex ratio, caused by infection of a population with a male-killer, may reduce female promiscuity and thus decrease the intensity of inter-male sperm competition. This would ease the passage of driving chromosomes. In contrast, a heavily female-biased sex ratio would lessen the intensity of intra-male sperm competition (the number of offspring sired is in direct relation to the number of sperm possessed); would select strongly for sperm number, and thus decrease the likelihood of spread of a driving chromosome. Again, this consideration is unlikely to be globally important, but may be a reality in drosophilids, where male-killers are known and meiotic drive is rather common.

Clutch size

Approximately half the eggs from a male-killed clutch fail to hatch. The invasion of a male-killer will clearly impose selection upon the size of egg clutch produced by the host. In the simplest model, we could imagine that a females clutch size decision was a trade-off between the cost (in time and energy) inherent in finding new sites to lay eggs, and the cost of avoiding excessive sibling–sibling competition amongst her progeny. In a population invaded by a male-killer, the optimum clutch size may change.

The nature of the change in clutch size will depend upon whether females can detect their status with respect to infection. If they can, then an increase in the clutch size produced by infected females only is likely to be favoured. An effect specific to infected individuals would mean that any reduction in antagonistic interactions between siblings in male-killed clutches (compared to normal clutches) will be lessened. The increase in clutch size, the response of the host, would therefore result in lower equilibrium frequencies of infection.

If females cannot detect their status with respect to infection (which will be true, at least initially), then selection may act to increase clutch size across all females. With a change in clutch size again comes a change in the level of antagonistic interactions suffered by siblings, and thus a change in the frequency of the male-killer. However, given antagonistic interactions are increasing across all clutches irrespective of infection status, the effect on male-killer infection rates is less clear, and will depend on the exact relationship between clutch size and neonate survival.

The idea that male-killers impose selection on clutch size evolution is one that can be assessed only after formal modelling and detailed experimental work.

Male-biasing agents and eusociality

Populations of haplodiploid species in which certain females produce just sons may predispose the population to the evolution of eusociality. As Godfray and Grafen (1988) note, it is sex-ratio heterogeneity that allows the spread of sibling helping genes, for in this context, help by females is directed more towards siblings of high relatedness (sisters) than low (brothers). It has been argued that the presence of *Wolbachia* in a population of Hymenoptera may produce the same effect (Hurst 1997). *Wolbachia* certainly is known in aculeates (Werren *et al.* 1995b), but what is yet to be determined is whether these *Wolbachia* cause CI. Beyond this, the presence of females producing only male offspring in a population (which will occur at equilibrium if *Wolbachia* infection does not go to fixation) will produce selection for modification of the fertilization rate, and thus the sex-ratio produced by individuals unaffected by *Wolbachia* action. Trivially, in a panmictic population the sex ratio these females produce is expected to become somewhat more female-biased, in accordance with Fisherian selection.

5.8.3.3 Male-killing symbionts and the population genetics of mitochondrial DNA

All inherited symbionts which spread from infection of a small number of host females cause an immediate decrease in the diversity of mtDNA. Rather than varying neutrally, the mtDNA type associated with the initial infection increases in frequency, hitch-hiking with the symbiont. This hitch-hiking has been observed for the cytoplasmic incompatibility (CI)-inducing *Wolbachia* in *Drosophila simulans* (Hale and Hoffmann 1990), and is expected for all symbionts which are spreading through selection.

The issue of mtDNA in species bearing a male-killing symbiont has recently been explored theoretically (Johnstone and Hurst 1996). A loss of diversity does result. The main difference between the effect of a male-killing symbiont and that of other symbionts is that a permanent, substantial loss of diversity may occur following the spread of a male-killer. Male-killing infections most commonly affect between 1 and 25 per cent of host females at equilibrium. This equilibrium is considered to be the product of losses through inefficient transmission balanced against a survival advantage to infected female hosts, as discussed above. At equilibrium, therefore, there is a constant movement of mtDNA variants from infected to uninfected females, through inefficient symbiont transmission. In the absence of horizontal transmission, mtDNA variants do not move in the opposite direction. Thus, the effective population size of mtDNA at equilibrium is lowered, to that of the size of the infected population, with concomitant decreases in standing mtDNA diversity. The same effects are expected for CI-causing *Wolbachia*; however, the high proportion of female hosts infected make the permanent reduction in mtDNA diversity very small in this case.

The selective sweeps on mtDNA by male-killers in particular, and symbionts in general, are not important in host adaptive evolution, except that such sweeps may allow the spread of mildly deleterious mitochondrial DNA mutants through hitch-hiking. However, they are important to evolutionary biologists who seek to use mtDNA as a tool in phylogeography. mtDNA diversity in symbiont-infected species does not behave as a neutral marker, and cannot be used to estimate population size or history. Lack of diversity of mtDNA does not necessarily indicate either a host bottleneck or continued small population size. It may be merely an indicator of infection with a symbiont. The history of mtDNA and nuclear genes will be very different in a species that is, or has been, infected with a male-killer.

5.9 Conclusions

The initial part of this chapter discussed a range of possible and observed distortions of the sex ratio by cytoplasmic factors. In two cases (cytoplasmic factors inducing meiotic drive in female heterogametic species and cytoplasmic elements preventing entry of X-bearing sperm into the ova in male heterogametic species), there was no definitive case study showing the existence of the trait. In one case (the alteration of fertilization frequency in arrhenotokous species) little is known about the nature or frequency of the elements. The initial focus for research here is to evaluate the frequency of the traits in question. Once this has been achieved, their role as influential passengers can be better evaluated. At the moment we have only hypotheses as to the importance of these sex-ratio distorters, in some cases without any evidence for there existence. For instance, the ability of cytoplasmic genes to influence segregation in species which are female heterogametic has been suggested as one reason why female heterogamety is rare, since it is a genetic system that is easily corrupted by inherited symbionts (Hamilton 1993). Although the idea is an attractive one, the lack of any good examples of the process envisaged must count against this hypothesis.

In the latter section of this chapter (male-killing), we reviewed a more comprehensive literature. We can safely say that early male-killing is common, at least in species with particular ecologies. Further, we understand much of the population biology of male-killing elements. Beyond this, we now know that this is a trait that has evolved many times in eubacteria, and this has raised the possibility that male-killing will evolve as a trait of all inherited bacteria where it is advantageous to the bacterium. This has led to speculation as to the incidence of different symbioses. In the case of male-killers, the stage is set for us to evaluate critically the idea that symbionts are important in host evolution. We have speculated as to the potential importance of male-killers in host evolution. We can now begin to perform empirical studies to test the predictions of these models. We will then be able to say with certainty whether male-killers really are influential passengers, or merely interesting curios.

6 The potential application of inherited symbiont systems to pest control

Steven P. Sinkins, Chris F. Curtis, and Scott L. O'Neill

6.1 Introduction

Arthropod pests can have an enormous impact on human life. Insect vectors are responsible for the spread of some of the world's major diseases, which in the tropics represent some of the most serious problems in health care. For example, the number of clinical cases of malaria per annum has been estimated at 350–450 million, including 1.4–2.6 million deaths (World Health Organization 1995). The impact of insect pests of agriculture is even more widespread. They can have dramatic effects on communities in developing countries when they are responsible for large yield losses in food crops, and are also of major significance in the developed world.

Chemical pesticides have long been the main weapon used for the control of insects, and have enjoyed considerable success. However, the widespread development of insecticide resistance has frustrated many attempts to control or eradicate serious pests. Very high costs are associated with both the development of new insecticides and with the increasingly heavy insecticide application needed to kill resistant pests. In many developing countries, where the need is greatest, resources are not available to mount effective chemical control programmes. Environmental safety concerns associated with insecticides such as DDT have also been a major stumbling block. In the case of vector-borne disease, the development of parasite drug resistance compounds these problems and increases the need for alternative, sustainable strategies for the control of the insect vectors.

Genetic control has the potential to be a powerful addition to the more traditional techniques of insect suppression. The elegant methods employed by inherited endosymbionts to distort host reproduction to their own ends are not only of interest in evolutionary terms, but also offer mechanisms which could prove extremely useful in this applied context. In fact, the phenomenon of cytoplasmic incompatibility has been the subject of speculation and experimentation with respect to the control of mosquito populations since the 1960s, predating the discovery of its endosymbiotic causation. The rapid advances now being made in DNA-based technology have renewed this interest and greatly expanded the range of possibilities for the utilization of inherited microorganisms, including such long-term goals as the creation and release of transgenic insects.

6.2 Endosymbiont distribution and interspecific transfer

A very desirable attribute for a genetic control system is the ability for it to be used in more than one species. It is a great advantage to be able to transfer techniques and experience developed in the control of one pest into that of another, and thus bypass the requirement for detailed molecular knowledge of each pest species. The recent discovery of the wide distribution across insect orders of very closely related *Wolbachia* endosymbionts (Chapter 1) suggests possibilities for the use of the CI mechanism across a range of pest species. *Wolbachia* are already known to occur in several disease vectors and agricultural pests; in fact PCR-based surveys of diverse insect species have indicated that naturally occurring infections may be present in more than 15 per cent of all insect species (Werren *et al.* 1995*b*).

The strong evidence for horizontal transfer between species which are only distantly related, based on non-congruence of host and bacterial phylogenies (O'Neill *et al.* 1992; Rousset *et al.* 1992*b*; Werren *et al.* 1995*a*), is a good indication that a particular strain of *Wolbachia* is not restricted to one host, and that the mechanism of CI must affect a relatively general and conserved target. This implies that transfers into previously uninfected host species should result in the expression of CI. Transfer of bacteria from *Aedes albopictus* mosquitoes into cured *Drosophila simulans* has already been achieved (Braig *et al.* 1994), and incompatibility was expressed when the transinfected male flies were crossed to uninfected females. In fact, a novel crossing type was generated: the introduced bacteria induced bidirectional CI with both the Riverside and Hawaii strains of *D. simulans*; Riverside and Hawaii being mutually incompatible strains (O'Neill and Karr 1990).

One could argue that, while it would be possible to move this agent between species which are naturally infected with *Wolbachia*, transfer into naturally uninfected species might not be successful because such species are resistant in some way to the action of CI, and hence do not already harbour the infection. However, given the obligate intracellular dependency of the bacterium, the main limiting factor in its current distribution may be the extreme rarity, on anything but an evolutionary time scale, of the natural establishment of germline infections in new species. The dynamics of CI dictate that if the infection lowers host fitness or if there is incomplete maternal inheritance then it will only spread if the number of infected individuals in a population is greater than a threshold level (Chapter 2); this further decreases the likelihood of natural horizontal transmission events. In short, it is very likely that many species have never harboured *Wolbachia* in their evolutionary history.

The artificial transfer of the bacterium into naturally uninfected species is therefore a theoretically attractive proposition, but in practice is technically far from easy. It is routine in *Drosophila* species to obtain high hatching rates after microinjection of embryos, but this is certainly not the case for most other insects. In mosquito species the survival rates after injection are generally very low (<5 per cent), and attempts to transfer *Wolbachia* into the naturally uninfected malaria vectors *Anopheles gambiae* and *An. stephensi* have not yet been successful. In order to achieve transfer of the bacterium into insects which are more 'delicate' than drosophilids, it may be necessary to develop more sophisticated methods to purify *Wolbachia* from other host tissues, while still keeping significant numbers of the symbiont alive. It remains possible that *Wolbachia* infections are unstable if moved into a very different host, with much lower rates of maternal transmission and, possibly, the activation of the host immune system. More extensive interspecific transfer in the lab is needed in order to investigate such effects. However, the strong evidence that *Wolbachia* has become established naturally, after interspecific transfer between insect families, should be cause for optimism.

The limited available evidence suggests that *Wolbachia* are somewhat exceptional in their wide distribution (Chapter 1), and other arthropod endosymbionts may not have such a wide host range. In the case of sex-ratio-distorting elements, the surprising degree of interspecific variation in sex determination systems may make it unlikely that any one endosymbiont species would be able to distort sex ratios in phylogenetically distant hosts (Chapter 5).

6.3 Cytoplasmic sex-ratio distortion in pest control

There are several ways in which endosymbionts could be utilized to suppress populations of insect pests. One possible application would be to enhance the use of predators or parasites of a particular pest species as biological control agents. A number of hymenopterans parasitize insects of agricultural importance, ovipositing in the pupae or eggs of the host insect and thereby leading to its death. Some are already commercially available as biological control agents, and obviously the extent to which they can be reared and released in sufficient numbers is critical to success. As only the females have a negative impact on the host insect, a genetic system which produced only females would halve the costs of any mass rearing programme. This doubling of the population size which is effective in control would continue to be seen in their offspring in the field. Symbiont-based sex-ratio distortion systems are ideal in this regard.

Parthenogenetic systems which result in all-female broods have been documented in parasitic wasps of the genera *Trichogramma* and *Muscidifurax*, and have been shown to be *Wolbachia* mediated (Rousset *et al.* 1992*b*; Stouthamer *et al.* 1993). *Muscidifurax* is a parasitoid of housefly pupae which has already been used in biological control trials (Petersen *et al.* 1992; Klunker 1994), while *Trichogramma* parasitizes the eggs of lepidopterans and has long been used against various pest species (Smith 1996). *Wolbachia* may be more widespread in its natural distribution within the Hymenoptera than has so far been reported (Zchori-Fein *et al.* 1995); in addition there is reason to expect that parthenogenesis may be induced in other hymenopteran species of commercial interest by artificial transfer of these *Wolbachia*. Sex-ratio distortion systems which are based on male-killing would be less readily utilized because they do not actually increase the number of females in a brood, although they may increase the survivorship of these females if they suffer reduced competition with their male siblings during their early growth (Hurst and Majerus 1993).

6.4 Cytoplasmic incompatibility and population suppression

The sterile insect technique (SIT), the mass release of males sterilized by irradiation to reduce the number of viable eggs being laid by wild females, has been the most successful application of genetic control to date. It was used from 1958 onwards to eradicate the screwworm fly (*Cochliomyia hominivorax*) from the USA, Mexico, and Libya (Krasfur *et al.* 1987; Lindquist *et al.* 1992). This was achieved despite relatively little knowledge of the genetics of screwworms (the SIT is regarded as a form of genetic control in the sense that sterility is caused by dominant lethal mutations in the gametes of irradiated males). SIT is also being applied with increasing success to medfly (*Ceratitis capitata*) control, but most attempts to transfer this technology to other species have not proven fruitful. It is a costly method which demands both considerable resources and a suitable life history in the target insect. One problem has been that males may suffer from reduced fitness compared to their wild competitors and suffer very low mating success, especially if the radiation dose required for sterilization is very high, as in lepidopterans (LaChance 1979). Therefore, a natural sterility-producing system without these high fitness costs would be very desirable in some cases. In fact *Wolbachia*-mediated cytoplasmic incompatibility fulfils these requirements.

Laven found that there are multiple crossing types in mosquitoes of the *Culex pipiens* complex, including the filariasis vector *C. quinquefasciatus* (Laven 1967b); in fact this mosquito group shows this phenomenon more extensively between populations than has been reported in any other insect. Laven predicted that populations could be controlled by swamping them with males of a strain with which they are bidirectionally incompatible, so that the wild females lay sterile clutches. The application of this principle was demonstrated in a WHO-sponsored field trial in Burma (Laven 1967*a*). The Burmese population had been found to be bidirectionally incompatible with a strain from Paris. The Paris mosquitoes themselves were not considered to be well adapted to Burmese conditions; therefore the genome of a strain from California which was compatible with the Paris strain was combined with the Paris cytoplasm by backcrossing. The ability to replace the genome of the insects to be released is an advantage rendered by the maternal inheritance of crossing type.* Laven targeted a small, relatively isolated mosquito population (in a village surrounded by dry fields), and by releasing his incompatible males in daily numbers comparable to the overall natural population size, 100 per cent of the wild egg rafts sampled became inviable within 12 weeks.

While this study was an impressive demonstration of the potential of CI for a release programme to control field populations, mosquitoes in the developing world are probably not a realistic target for such strategies. Their high population densities and population recovery potential, coupled with their wide distribution over very large areas, would render any population eradication scheme based on sterile release unaffordable except on a local scale. Relatively small-scale local releases would necessitate a continual battle against population re-establishment through immigration, a battle which was carried on

---

* In fact, in subsequent work in India the genome of the target population itself was introduced, maximizing the fitness of the males to be released, via a 'bridging strain' compatible with both (Krishnamurthy and Laven 1976; Curtis *et al.* 1982). This can also now be achieved by simply curing wild-type males of their *Wolbachia* using tetracycline.

for many years with screwworm fly sterile male releases in Texas and northern Mexico. The biology and life history of a species, especially its potential reproductive rate and population regulation by density-dependent factors, are crucial factors with respect to the effectiveness of the release of incompatible males, and thus its feasibility as a control measure.

Mass male release is unlikely to be an economically viable control strategy for many vectors of human disease in developing countries, which are often already under pressure to cut their public sector health budgets. This would not present such a problem for agricultural control programmes. Insect pests of cash crops, such as cotton, fruit, coffee, or cattle, are associated with tangible cash losses, against which the high cost of a release programme may be offset. Agricultural insect pests in developed or partially developed countries which have the necessary infrastructure for releases over wide areas probably represent the most realistic targets for CI-based suppression strategies.

One disease vector whose unique biology has attracted several attempts at sterile male control are tsetse flies (*Glossina* spp.), which transmit the African trypanosomes responsible for human sleeping sickness and nagana in cattle. They have a very low reproductive rate, producing only one well-developed larva every 9 days, so that recovery after population suppression is relatively slow. This life history also renders mass rearing rather difficult, although several SIT programmes (e.g. in Zanzibar) have already been undertaken which demonstrate that these difficulties are not insurmountable if suitable funding is made available. SIT in tsetse is complicated by the fact that both male and female flies take blood meals and therefore transmit trypanosomes. Therefore, released sterile males have the potential to increase the disease burden in a given area. The current SIT programme that is under way in Zanzibar, against *G. austeni* which do not bite humans, is overcoming this problem by feeding males on a blood meal supplemented with anti-trypansomal drugs prior to release (U. Feldman, personal communication). In addition, irradiated males have reduced life spans which limit their potential as disease transmitters.

Some populations of tsetse fly are known to harbour *Wolbachia* (O'Neill *et al.* 1993). Therefore CI could be very useful in their control, perhaps to complement the SIT programme. If the irradiated males being released are also cytoplasmically incompatible with the target population, this would allow a significant reduction in the radiation dosage, and hence an increase in male competitive ability. The combination of irradiation with CI has been investigated in *Culex* (Arunachalam and Curtis 1985). Such a release strain of tsetse would also need to be refractory to trypanosome transmission (unable to transmit the parasite) to maximize the effectiveness of such a strategy. This refractoriness could be achieved in a number of ways, including the feeding of anti-parasitic chemicals to the flies during rearing, selecting fly strains which are refractory, or genetically engineering a strain of refractory flies. The latter option has become feasible in recent years now that a method is available to indirectly transform tsetse via their midgut symbionts (Beard *et al.* 1993).

Despite extensive studies in the 1940s (Vanderplank 1947), 1960s (Curtis 1972), and 1980s (Gooding 1990), crossing patterns between populations of tsetse, and their relationship to the presence or type of *Wolbachia*, are still inadequately known. Crossing surveys to determine natural crossing types will first need to be performed if the target populations are naturally infected with *Wolbachia*. Where the target population is uninfected, males could be reared and released from any infected population which is able to induce unidirectional incompatibility with the uninfected flies, provided that there are no behavioural barriers to cross-mating in the field. *Wolbachia* does not have to be present naturally in a pest species for CI-based sterile release to be employed, if the bacterium can be artificially introduced and CI is expressed in crosses with uninfected populations.

In general, a CI-based sterility programme would depend heavily on the use of an efficient method of sexing for the removal of females. Any released females which had not been sterilized by an alternative mechanism would be fertile if they mated with the released males, or with both wild and released males if the incompatibility with the target population was only unidirectional. Removal of females would be, in any case, an important part of most SIT programmes, as even sterile females may transmit disease or cause oviposition damage to soft fruit. Ideally, sexing systems would be genetically based to reduce costs, such as the translocation on to the Y chromosome of pupal colour mutations in medfly (Robinson and Van Heemert 1982; McInnis *et al.* 1994) or insecticide resistance in *Anopheles* (Curtis *et al.* 1976).

The potential for the use of CI in this manner would need to be evaluated in each individual species because of the problem of reduction in the penetrance of incompatibility with male age (Singh *et al.* 1976). This decline shows a great deal of variation between species, for example being very rapid in *Drosophila melanogaster* (unpublished data) but much less so in the closely related species *D. simulans* (Hoffmann and Turelli 1988; Turelli and Hoffmann 1995). Direct assessment of this effect in the target species, using crossing experiments, would be necessary.

6.5 Pest population replacement and bidirectional CI

The goal of almost every pest control strategy which has been attempted, from insecticides to SIT, has ultimately been to reduce the pest population size to as great an extent as possible. However, there are some fundamental disadvantages to this approach, in whatever form it takes. A reduction in population size will generally encourage population recovery due to density-dependent reduction of competition, so that survivors and immigrants are at a reproductive advantage (Curtis and Graves 1988). The problem would only cease if the species could be completely eradicated, which is an unrealistic goal for most pests. It has therefore been proposed that a more sustainable approach to control, especially of insects which transmit diseases, would be to render the pest population harmless rather than attempting to eradicate it. The aim would be to introduce useful genes into populations, for example to render them unable to transmit a pathogen to humans, agricultural animals, or plants.

A principle which applies to any genetic system of cross-sterility between insect strains in sympatry is that the females of the minority strain would be subject to a greater reproductive disadvantage than would females of the majority strain, due to the preponderance of the males with which they are unable to interbreed successfully (Curtis 1968). Therefore the theoretical prediction is that the minority strain would be replaced by the majority strain in the absence of any other selective forces: a stable equilibrium between such sympatric strains could not exist. An important cross-sterility system to which this principle would apply is bidirectional incompatibility in *Culex pipiens*.

If both sexes of a bidirectionally incompatible strain were introduced so as to form a population majority, with the aim of replacing the local wild type (Laven and Aslamkhan 1970), then the nuclear genome of the released mosquitoes would be expected to spread in association with its cytoplasmic type, as long as there is complete sterility in all cross-matings. The newly established released strain should be able to sterilize immigrants of the original crossing type and thus prevent re-establishment of that type. With this system the requirement for sterility is more stringent than would be the case for the use of bidirectional CI in an SIT programme, because of the necessity that the useful gene remains 'linked' to the introduced cytoplasm.

Trials to test this strategy were carried out in laboratory cage populations of *Culex quinquefasciatus* (Curtis and Adak 1974) and in population field cages in India (Curtis 1976). Although the foreign cytoplasm did indeed replace the wild type, linkage with a semi-sterilizing male-linked translocation was not fully maintained. It appeared that effects such as reduced incompatibility with male ageing (Singh *et al.* 1976; Subbarao *et al.* 1977*a*) and segregation of crossing types within populations (Subbarao *et al.* 1974, 1977*b*) had caused a breakdown in absolute sterility between the populations. The resulting loss of linkage between the nuclear genotype of interest and the cytotype being released would be fatal to an aim of complete population replacement. The effects of ageing on CI do show variation between species, and it is possible that the potential for loss of linkage would not be an insurmountable problem in other systems.

The great advantage of genetic replacement using bidirectional CI is that it does not require any molecular knowledge of the genes which actually render the pest species harmless. Classical selection procedures could be employed for a particular trait, such as the inability to transmit a disease agent. The genetics of susceptibility to pathogens has been investigated in several vector insects; the most important example is heritable malaria refractoriness in *Anopheles* mosquitoes. Several examples which are selectable from naturally occurring populations have already been uncovered (summarized by Curtis 1994), some of which are based on encapsulation of the oocysts of the parasite. In theory, if *Wolbachia* could be introduced into these mosquitoes and bidirectional CI generated, naturally susceptible populations could be replaced by the release of refractory individuals. However, it is very doubtful that the mass release of fertile malaria-refractory females would be an acceptable strategy, as they would still be biting humans. This would not only add to the mosquito nuisance problem, they would also be capable of transmitting other diseases, such as filariasis and some arboviral infections. In addition, although the released females should not actually function as malaria transmitters, their descendants potentially could do so if there was some degree of fertile mating with the wild males.

Many of these concerns would be removed if the host-biting behaviour itself were the target of manipulation. The most important malaria vector, *An. gambiae* s.s., is attracted to human odours, while its sibling species *An. quadriannulatus* is not. As fertile females are produced when the species are crossed, it might prove possible to replace the genes which are presumably responsible for this behavioural difference in *An. gambiae*. The success of any attempt to spread such genes would be dependent on the fitness cost imposed by blocking the attraction to humans, determined by the degree to which the female mosquito would still be able to locate other hosts (such as cattle) and take blood meals from them.

There are severe practical constraints associated with the use of bidirectional CI for pest replacement, in human disease vectors at least. They rely on large-scale release, although not on the same scale as an SIT programme since only a small majority of released insects need be achieved in the population for a short time. The likelihood of success could be enhanced by releasing into wild populations at times of seasonal population lows or after an insecticide campaign, but it would still be an expensive strategy.

6.6 Unidirectional CI as a gene transport mechanism

What is needed to bring population replacement strategies within the range of economic feasibility for a control programme in the developing world is a self-driving system. It must be able to spread useful genes through a population from a relatively small population seeding, rather than having to release such numbers that a local majority was achieved.

Unidirectional CI between infected and uninfected individuals, the mechanism that the bacterium itself employs to invade a naïve population, provides just such a self-spreading mechanism. The invasion of *Drosophila simulans* in California is a demonstration of the power which such a system possesses to invade natural populations (Turelli and Hoffmann 1991). The infection wave has been reported to spread geographically at a rate of around 100 km/year, and reached fixation within a relatively small number of generations (Chapter 2). There is a price to be paid for this spreading power, however. Unidirectional CI could not be used to spread chromosomally located genes, because the fertile mating between infected females and uninfected wild males would allow wild-type genes to be recombined quickly with the genome of the introduced *Wolbachia*-infected insects.

The dynamics of the uniparental inheritance of the cytoplasm, where no recombination can occur because the male makes no contribution to inheritance, dictate that any maternally inherited gene would be spread by a *Wolbachia*-mediated unidirectional CI system. It is unlikely that many naturally occurring genetic systems which would render the pest harmless will prove to be maternally inherited, although it should be noted that in tsetse flies, maternally inherited symbionts have been implicated in vector competence with respect to trypanosome transmission (Maudlin 1991). Therefore the use of unidirectional CI introduces a requirement that the genes responsible for rendering the pest harmless must be relocated in a maternally inherited expression vector, and thus their molecular isolation becomes essential. It would also rule out the use of a polygenic system, i.e. monofactorial inheritance is required. In addition, the gene will have to be expressed in a wild-type background, and must therefore be dominant in its effect relative to any wild-type alleles (for example those causing susceptibility to a disease agent). The use of unidirectional CI as a gene transport system therefore demands the use of recombinant DNA technology to create transgenic insects.

Any species or population which does not currently harbour *Wolbachia* represents a target for population replacement using unidirectional CI. If *Wolbachia* can be interspecifically transferred into an uninfected pest species, and if this results in the expression of CI, the spreading process should occur as in *Drosophila*. Unidirectional CI has also been shown to occur between different populations which are both infected with *Wolbachia* in *Culex* spp. (Yen and Barr 1974), *Aedes albopictus* (Kambhampati *et al.* 1993), *Drosophila simulans* (Nigro 1991), and *Nasonia* wasps (Breeuwer and Werren 1993). The dynamics of unidirectional CI between infected populations produce the same effect as that seen between infected and uninfected insects. That is, one of the *Wolbachia* types will replace the other, just as the infected state will replace the uninfected. If this phenomenon can be suitably manipulated, species and populations which are naturally infected with *Wolbachia* would also be targets for population invasion.

As already mentioned, *Wolbachia* which cause fitness reduction or which are imperfectly transmitted need to reach a threshold frequency within a population before they will spread to fixation. The rate of spread of *Wolbachia* within a population will depend on the natural mobility of individuals and the degree to which their movements are influenced by human transportation, which will obviously vary between species. *Drosophila* are closely associated with human activity and are known to be spread with the transportation of fruit, and many pest insects are likely to show the same kind of mobility as *Drosophila*. However, if the species population structure consists of numerous fragmented sub-populations, which are relatively isolated with little inter-migration and gene flow, this would impede the successful spread of the bacterium from a small-scale local release.

There is generally not a great deal of information available about population structure which would allow these important considerations to be evaluated with any confidence in a particular species. However, circumstantial evidence does exist that extensive gene flow may occur in mosquitoes. In *Culex pipiens*, amplified B2 esterase genes, which render the insect resistant to organophosphate pesticides, occur in many places of the world. It has been argued that the mutation has occurred once and then spread extensively, based on high conservation of the gene in different geographical populations (Raymond *et al.* 1991). This would indicate that extensive migration and inter-population dispersal can occur in this species for a gene subject to strong selective pressure.

6.7 Modification of pest populations

Three objectives must therefore be achieved if transgenic insect pests are to be used to modify field populations. First, a dominant gene rendering the insect harmless, or at least reducing its impact as a pest, must be identified and cloned. Secondly, it is necessary to introduce and express such a gene in the individual, in such a manner (temporally and spatially) that it has the desired effect. Thirdly, this expression system must be capable of spreading or being spread through natural populations, probably to fixation. It is very important that the three lines of research should proceed in synchrony, as they are closely interrelated, and each will influence the range of possibilities that would be acceptable in the other two.

6.7.1 Useful genes in pest control

The most obvious genes with which to modify a vector population are those which naturally render some individuals or populations refractory to the transmission of a pathogen. If natural refractoriness mechanisms are to be spread by a unidirectional CI system, the identification and cloning of the genes responsible is essential. Genetic maps are being constructed and applied to the identification of refractoriness genes in *Anopheles gambiae* (Zheng *et al.* 1993), although such maps are not available for most insect species, and their construction is a laborious undertaking. A second limitation is that the mechanism would need to be genetically dominant and controlled by one or a very small number of genes, which would limit the number of examples which could be employed in this way.

Another approach which has been proposed is to use foreign gene products rather than manipulating naturally occurring refractoriness mechanisms. In this way the difficulties associated with the isolation of the host genes controlling the process are bypassed, and therefore the applicability of the strategy is broadened. Secondly, the requirement for a genetically dominant, relatively small construct is satisfied, which often would not be the case for naturally occurring refractoriness genes. The mammalian antibody-based immune system is an obvious source of foreign gene products which could exert an antiparasitic effect if expressed in insects. The parasite stages found in the insect will not have been subjected to antibodies in their evolutionary history; they would therefore be expected to be more amenable targets to antibody attack than would the antigens of the mammalian stages of the parasite. Monoclonal antibodies have already been raised against mosquito-stage antigens of *Plasmodium* and they block malaria transmission efficiently (Winger *et al.* 1987). A gene expressing a single-chain form of such an antibody could be expressed in the insect in order to reduce its competence as a vector.

In fact any arthropod-transmitted disease agent could, in theory, be amenable to this kind of approach if antibodies can be raised against suitable antigens. In the case of viruses, antisense RNA products, which would interfere with replication, constitute a second class of artificially constructed antiviral gene products being investigated (Olson *et al.* 1996). These approaches are not restricted to pathogens of humans and agricultural animals, but could also be applied to disease agents of crop plants. Much progress is already being made with the introduction of foreign genes in plants, such as the expression of animal antibody constructs against viruses in tobacco (Taviadoraki *et al.* 1993). In the case of crops used for human foods, however, it may prove a more desirable goal to introduce foreign antiparasitic genes into their insect vectors rather than into the plants themselves, thus avoiding issues of consumer safety or confidence.

Another class of useful genes are those which would increase the efficiency of population suppression strategies in integrated control programmes, particularly the combination of chemical and biological control. For example, alleles which increase susceptibility to a particular pesticide might be spread by a *Wolbachia*-based system, followed by application of chemical control. This could represent a means of managing the widespread problem of the development of insecticide resistance. However, most examples of insecticide susceptibility which have been investigated are partially or completely genetically recessive, and would therefore require genetic homozygosity if susceptibility is to be restored. There are exceptions to this generalization, including most examples of susceptibility to pyrethroids (Farnham 1977; Priester and Georghiou 1979; Lin *et al.* 1981; Curtis *et al.* 1993). It is conceivable that other such dominant genes conferring susceptibility to other insecticides will be isolated.

The danger inherent in any strategy to interrupt the transmission of a pathogen by the insect vector is that a selective pressure will be imposed on the parasite to evade this process. The widespread evolution of drug resistance in parasites such as *Plasmodium* provides a sobering demonstration of their evolutionary plasticity. It is to be hoped that the use of recombinant antibodies against antigens which are not normally exposed to them, and therefore show lower levels of variation, would reduce this risk somewhat. If two independently acting genes could be expressed, then a mutation in the parasite which allowed it to evade one of the mechanisms would still not be able to spread because of its vulnerability to the other. The risk of the development of pathogens able to evade both mechanisms should therefore be very low.

6.7.2 Gene expression systems

For the spread of useful genes by unidirectional CI, the most obvious candidate for transformation as an expression vector is *Wolbachia* itself. However, the location of *Wolbachia* in the host presents a fundamental problem to this strategy. Studies in *Culex* have indicated that it is restricted to gonadal tissues only (but tissue distribution has not been widely studied in other species). If it is restricted to the germline tissues in most insects then its usefulness is constrained, given that the expressed gene products would generally have to be active in tissues such as the midgut or salivary glands to have antipathogenic action in disease vectors, or at least be exported into the haemocoel. Export of gene products from the ovaries should not be impossible; these tissues are, after all, the site of manufacture of hormonal proteins which have sites of action all over the body of the insect. One case, in which the gonadal location of *Wolbachia* would not be a disadvantage, would be if the target pathogens are themselves transmitted through the reproductive tissues of the arthropod, for example arboviruses such as La Crosse and also the rickettsial disease agents. Interrupting this transovarial transmission in the reproductive tissues of their arthropod host could have great impact on the normal ecological cycle maintaining these disease agents.

We have already seen that any maternally inherited genetic element would be spread by an invasion of *Wolbachia*. Some other such elements have a much more appropriate tissue distribution for the expression of antipathogenic agents than does *Wolbachia* itself, and could therefore represent more suitable candidates for gene expression. The invasion of *D. simulans* populations in California by *Wolbachia* was associated with the spread of a mitochondrial variant, in keeping with theoretical predictions for the population dynamics of cytoplasmic inheritance (Turelli *et al.* 1992). The mitochondrion is an ideal example of a maternally inherited element present in all insect tissues cells. In fact mitochondria have been transformed in yeast, although this was aided by the fact that yeast cells are able to survive in the absence of mitochondria, which means that deficient cells could therefore be easily transformed and selected. To transform insect mitochondria would be a technically difficult task, although it is by no means theoretically impossible and may eventually be achieved given the rapid advances being made in recombinant DNA technology.

There are a number of nutritive bacterial symbionts which are also maternally inherited and are therefore candidates for use as gene expression systems, if they can be transformed. The first demonstration of transformation of this nature was performed in the bacterium *Rhodococcus rhodnii*, an endosymbiont found in *Rhodnius prolixus* (the bug which transmits the trypanosomes causing Chagas disease). The bacterium was transformed with an antibiotic resistance gene (Beard *et al.* 1992), using a shuttle vector plasmid containing an origin of replication isolated from an endogenous plasmid. Recent studies have shown that these symbionts are capable of expressing introduced cecropin constructs which can offer protection against trypanosome infection (Durvasula *et al.*, 1997). However, *R. rhodnii* is an extracellular symbiont spread by coprophagy rather than by maternal inheritance, and therefore would not be compatible with a *Wolbachia*-based population spreading system.

Tsetse flies have been shown to harbour three different bacterial endosymbionts (O'Neill *et al.* 1993; Aksoy 1995; Aksoy *et al.* 1995). A large primary symbiont (*Wigglesworthia glossinidia*) inhabits the midgut-associated mycetomes, while a smaller 'secondary' symbiont inhabits midgut epithelial cells. Both of these symbionts belong to the γ-Proteobacteria. The third intracellular prokaryote associated with tsetse flies is *Wolbachia*, located in the gonadal tissues. The primary and secondary symbionts are apparently inherited through the milk-gland secretions of the uterus. Although this is very different to the transovarial route of inheritance taken by *Wolbachia*, they would still be spread by an invasion of *Wolbachia* because their inheritance is solely maternal.

The tsetse secondary symbionts have been cultured and transformed with a plasmid containing an antibiotic resistance gene as a selectable marker, using an origin of replication of broad host range (Beard *et al.* 1993). There is a great deal of potential in this system for the use of gut-tissue endosymbionts as expression systems for products which attack trypanosome parasites, which could then be driven into populations by *Wolbachia*. Although examples of other nutritive symbiotic bacteria are not uncommon in insects, they do appear to be primarily associated with those species which have a restricted diet (such as those which feed solely on blood) and require nutritional supplements which the symbionts supply. They have not been found in mosquitoes, which may be a consequence of the wider dietary range of these insects.

A second category of maternally inherited elements which might be used as expression vectors are transovarially transmitted viruses. The best documented of these is the sigma virus of *Drosophila melanogaster*, reviewed by Fleuriet (1988). The only known mode of transmission of the virus is by vertical inheritance, which occurs at very high rates in stabilized infections (Fleuriet 1982); non-stabilized infections with lower transmission frequency can also occur, for example if only the male parent is infected. Sigma could, in theory, be spread through a population by an invading *Wolbachia* infection. Viruses in fact are already being used for transient expression of genes in mosquitoes (Carlson *et al.* 1995). The same techniques applied to inherited viruses could represent a novel means for population transformation.

It is very likely that many such viruses remain undiscovered, but there is a real difficulty in locating their presence, given that they may not have any phenotypic effects on the host. Sigma virus was only described because of its curious property of inducing CO_2 sensitivity in the host, i.e. paralysis of infected flies after exposure to this anaesthetic. Other rhabdoviruses such as vesicular stomatitis virus (VSV) also induce CO_2 sensitivity in mosquitoes (Rosen 1980). In fact this property is not restricted to rhabdoviruses: in certain populations of *Culex* mosquitoes CO_2 sensitivity was found to be induced by an inherited virus of unusual morphology called Matsu, which has not been phylogenetically placed (Shroyer and Rosen 1983; Vazeille *et al.* 1992). However, it could be expected that a number of maternally inherited viruses have no phenotypic effects at all, and therefore have not been described.

The great danger inherent in using *Wolbachia* to spread a separate inherited gene expression system is that the two will become dissociated, and thus the useful gene separated from its intended driving system. The most important factor to consider in this regard is the rate of maternal inheritance of the expression vector, which will determine the percentage of offspring which inherit *Wolbachia* but not the expression system itself. In practice, if the rate of maternal inheritance of the expression vector was significantly lower than 100 per cent, a complete population transformation with the useful gene would no longer be possible. Bacterial nutritive symbionts show very high rates of maternal transmission, and in addition any progeny which did not become infected would die if the relationship is obligate for the host, or be subjected to reproductive sterility (in the case of tsetse) or reduced growth rates. Therefore the risk of the spread of a *Wolbachia* infection no longer associated with the expression vector would be slight or non-existent in the case of nutritive symbiotic bacteria, which is a factor acting in their favour. The restricted phylogenetic distribution of inherited nutritive symbionts is the most significant limitation to their use.

However, the imperfect maternal transmission of inherited viruses presents more of a problem. Even stabilized sigma lines in the laboratory give rise to small numbers of uninfected insects, so some degree of dissociation must be regarded as inevitable. The degree to which dissociation occurs early on in the spreading process would be critical to the final population frequency reached by the virus. The danger would be increased if the virus imposes an additional fitness cost on the host, which would impart some selective advantage to those infections of *Wolbachia* which were no longer associated with the virus, and thus increase their rate of spread relative to combined infections. Transovarially transmitted viruses are also inherited through the male to some degree (e.g. sigma virus). In fact this is favourable to the spread of the virus, and may still allow a high percentage of the insect population to become infected with the transformed virus, even if some degree of dissociation with *Wolbachia* is occurring during its spread.

Wolbachia transmission through the male has been observed at frequencies around 1 per cent in laboratory *Drosophila* populations (Hoffmann and Turelli 1988). Paternal transmission of nutritive bacterial symbionts, on the other hand, is unlikely to occur because, in most cases, they are absent from the testicular tissues. Therefore this could represent another means by which the two could become dissociated, if *Wolbachia* alone is passed on through the sperm. However, from the evidence of mitochondrial DNA typing, it appears that paternal inheritance of *Wolbachia* is extremely uncommon in the field (Turelli *et al.* 1992), probably due to the lower densities of the bacterium which seem to occur in nature. Therefore the risk of dissociation imposed by paternal inheritance of *Wolbachia* is probably very slight. It is even more unlikely that a male infected with both *Wolbachia* and an inherited virus would transmit only the *Wolbachia* through its sperm.

Bacterial and viral genomes in general are smaller, simpler, and easier to manipulate than those of eukaryotes. For many pests, they may prove to be more amenable goals for transformation than the nuclear genome of the insect. These systems could also prove to be of use to express useful genes in beneficial arthropods which are mass released for the biological control of pests (and do not therefore require any gene transport system). An example of a trait which would be very useful to induce in this way is insecticide resistance in pest predators (Hoy 1990). The main disadvantage of symbiotic expression systems relative to nuclear expression is their limited distribution within host tissues and the need for the export of the products of introduced genes to their site of action. Furthermore, the use of prokaryotes to express foreign genes would probably preclude the utilization of insect gene promoters to temporally limit the expression of the useful gene, such as those which induce gut-specific expression after a blood meal (Muller *et al.* 1993).

6.7.3 Repeated population sweeps with *Wolbachia*

As already emphasized, there are two important areas where any strategy to spread useful genes through disease vector populations would be at risk of failure in the field: the potential dissociation between a useful gene and its driving system, and the potential development by the parasite of the capacity to evade the refractoriness mechanism. If a population sweep does not achieve complete success in rendering a population harmless, the ability to be able to induce one or more further invasions of the same population would be vital. The occurrence of unidirectional CI between infected populations provides a mechanism not only for the invasion of naturally infected populations, but also for repeated sweeps into naïve populations. If both *Wolbachia* types produce CI when infected males are crossed to uninfected females, it would be possible for a naïve population to be invaded first by one *Wolbachia* type, and then subsequently by a second type which was unidirectionally incompatible with the first.

Unidirectional CI between infected strains has been set up artificially in *Drosophila simulans* by using microinjection techniques to combine two mutually incompatible strains of *Wolbachia* in one host (Sinkins *et al.* 1995*b*). Males carrying both strains were incompatible when crossed to females containing only one of these same *Wolbachia* strains (Fig. 6.1), but the reverse cross was compatible. This situation is somewhat analogous to that seen with the ABO human blood group system, where the presence or absence of the A or B antigenic alleles determines the compatibility of donors and recipients. In the same way, the presence of two strains of the bacterium, which have presumably independent molecular mechanisms of inducing CI, results in incompatibility with females containing only one of the strains. At the population level the dynamics of a mixed single/double infected population (modelled in Chapter 2) mirror those of an infected/uninfected mixed population, i.e. double infections should replace single infections.

Double infections have also been found in natural populations of *D. simulans* and appear to have a similar effect on crossing type (Rousset and Solignac 1995). *Aedes albopictus* is an example of a mosquito species in which there is good evidence that some populations are naturally superinfected with different *Wolbachia* strains (Sinkins *et al.* 1995*b*; Werren *et al.* 1995*b*), and this is consistent with the predicted pattern of incompatibility expected between superinfected males and singly infected females. An alternative explanation of unidirectional CI between naturally *Wolbachia*-infected populations is that variations in bacterial density between populations are responsible for the

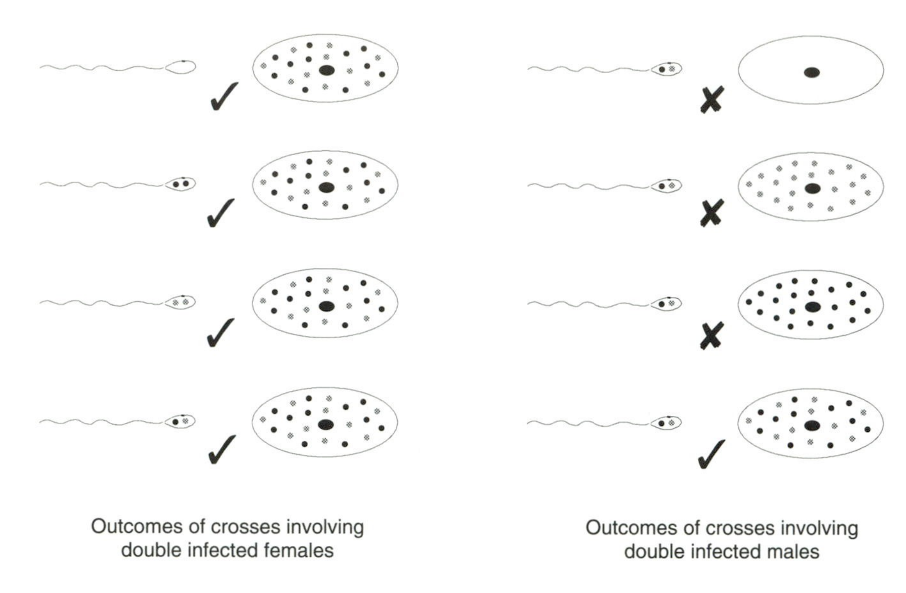

Figure 6.1 Crossing types generated when individuals with multiple *Wolbachia* infections are crossed. In this case the two *Wolbachia* strains are bidirectionally incompatible with each other.

different crossing types observed (Breeuwer and Werren 1993; Sinkins *et al.* 1995*a*). More work is needed to distinguish these effects.

It should be possible to induce a population sweep through a naïve population using a particular strain of *Wolbachia*, and subsequently a second invasion of the same population with a two-strain cytotype. Potentially, a third strain, mutually incompatible with the others, could be used to effect a further population replacement; the limits to the number of strains which could stably coexist are yet to be investigated. Some degree of segregation of singly infected individuals was seen in laboratory superinfected *D. simulans* (Sinkins *et al.* 1995*b*) and would be expected in natural populations during the spread of the superinfection. This segregation would be likely to increase as the number of strains increased, and probably would impose limits on the number of invasions of the same population which could be achieved. A preferable approach would be to combine the CI genes from mutually incompatible strains in the same bacterium; the possibility of segregation of single-infected offspring of a host infected with combinations of *Wolbachia* strains would thereby be avoided.

6.8 The isolation of incompatibility genes

If the genes that cause CI could be isolated, transferred on to insect chromosomes, and expressed appropriately so as to induce incompatibility, then would they spread into an insect population? If we could express CI genes appropriately from a chromosomal location, then we would expect that the patterns of incompatibility generated between various crosses would appear as outlined in Table 6.1. The central assumptions of this model are that allele A (which represents the presence of the CI gene(s)) is completely dominant with respect to imprinting sperm in males and rescuing sperm in eggs, and with respect to the effect on female fecundity. Based on these assumptions the frequency of these genotypes in an interbreeding population can be modelled as follows:

Let P denote the frequency of AA adults; Q, the frequency of Aa adults; $R = 1-(P+Q)$ = frequency of aa adults; F, the relative fecundity of individuals carrying A (aa individuals are assigned a value of 1); H, the relative hatch rates from incompatible versus compatible crosses; $sh = 1-H$ and $sf = 1-F$.

Set:

$$\overline{W} = 1 - sf(1-R) - shR(1-R),$$

then the recursions for the genotype frequencies of adults are:

$$\overline{W}P' = F\,[P(P+Q) + (1/4)\,Q^2]$$

$$\overline{W}Q' = (1/2)\,PQF + FPR + (1/2)\,FQ + HPR + (1/2)HQR.$$

This model predicts an unstable equilibrium point, which is sensitive to fitness costs associated with carrying the CI genes (Fig. 6.2). Above this equilibrium the CI genes would spread into an interbreeding population, below it they would decrease in frequency.

In theory, CI genes themselves, rather than *Wolbachia*, could be used as a tool to drive chromosomal genes into populations. The modelling shows that

Table 6.1 Assumed viabilities for different crosses involving individual insects carrying CI gene(s) on a chromosome and expressed appropriately. The egg cytotype represents the presence of CI 'rescue' factors (CI) in the egg or their absence (–).

Egg cytotype	Egg genotype	Viability
CI	AA	✓
CI	Aa	✓
CI	aa	✓
–	Aa	✖
–	aa	✖[a] / ✓

a Die if father was Aa.

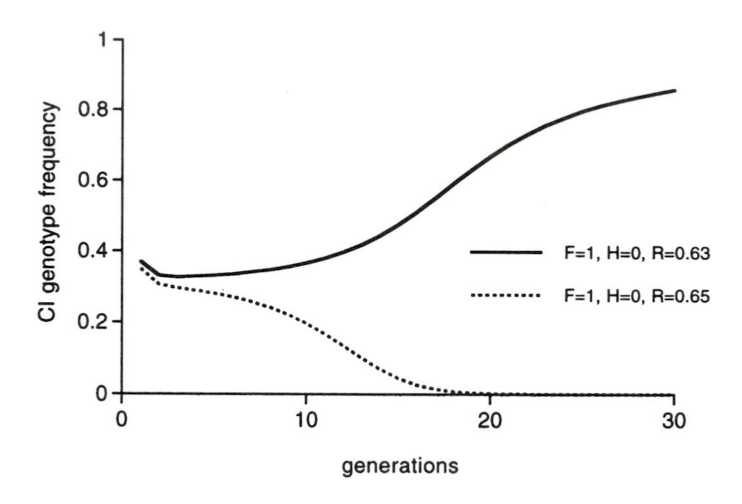

Figure 6.2 Dynamics of CI genes when located on insect chromosomes, showing an unstable equilibrium point above which the CI genotype will increase in frequency and below which it will be lost from a population.

this approach is unlikely to be as strong a driving system as the maternally inherited *Wolbachia* organisms themselves (Chapter 2). The threshold gene frequency that needs to be exceeded would require greater release numbers than a release involving a natural infection, and the speed in which the spread would occur would be slower than natural CI. However, this system would potentially obviate the biggest disadvantage of natural *Wolbachia*-mediated CI as a gene driving system, namely the requirement for a maternally inherited gene expression system to be linked to it. Another advantage would be the possibility of using combinations of CI gene sets from different *Wolbachia* strains in order to mimic superinfection spreading. This might allow for repeated population sweeps without the constraint of a host carrying a limited number of competing bacterial strains.

There are several additional reasons why it would be beneficial to isolate the genes controlling CI, besides the obvious importance of understanding the molecular basis of CI. For example, it should make it possible to separate *Wolbachia* strains which induce incompatibility from those which do not, without needing to perform crossing experiments. In some insects crosses are very difficult or impossible to carry out, especially in those cases where the insect cannot be cultured in the laboratory. It is possible that it would also allow prediction of which strains would be mutually incompatible. As well as facilitating the use of bidirectional CI, this knowledge would increase our understanding of how to use mutually incompatible strains for repeated sweeps through populations.

6.9 Conclusions and prospects

Given the current rate of advance in recombinant DNA technology, it seems likely that these techniques will have an increasing role to play in pest control. Endosymbiont systems offer both a route to gene expression, and a potentially powerful population-level spreading system in CI. Much more needs to be known about these systems, though, if such technically demanding tasks are to be achieved. There is also much work to be done on the isolation of suitable genes to render pests harmless.

Clearly, many years of research are required before transgenic insect pests will be ready for release into the environment. Nevertheless, it is never too early to give consideration to the possible implications of such releases. It would be important to plan any field trials using unidirectional CI very carefully, and not rush the commencement of an irrevocable release programme, to avoid wasting population sweeps which presumably are limited in number. It would be even more important to ensure that the released insects did not actually worsen the pest problem, by such means as unforeseen enhanced transmission of other pathogens, although any such effects must be considered unlikely. Even though the potential risks associated with a population replacement programme of this nature would probably be minimal, any release of pests into the environment for the purpose of genetic control would probably attract controversy, based on fear of the unknown. This was seen in past mosquito release trials in India (Curtis and Von Borstel 1978). It would be important to ensure that accurate information about the considerable potential benefits of this research is always available to the public.

The experimental and theoretical evidence collected to date indicates that symbiont systems offer considerable potential for pest replacement strategies in the future. Meanwhile, their use for the purposes of sex-ratio distortion in the production of biological control agents and in sterile insect release may have significant utility in the shorter term.

Acknowledgements

The authors would like to thank Ary Hoffmann, Michael Turelli and Jack Werren for comments on earlier versions of this paper. We are especially grateful to Michael Turrelli for his contributions in developing the model presented in section 6.8. This work was financially supported by NIH, WHO/TDR, and the McKnight foundation.

References

Adams, J., Greenwood, P., and Naylor, C. (1987). Evolutionary aspects of environmental sex determination. *International Journal of Invertebrate Reproduction and Development*, **11**, 123–136.

Agarwala, B. K. (1991). Why do ladybirds (Coleoptera: Coccinellidae) cannibalize? *Journal of Bioscience*, **16**, 103–109.

Aksoy, S. (1995). *Wigglesworthia* gen. nov. and *Wigglesworthia glossinidia* sp. nov., taxa consisting of the mycetocyte-associated, primary endosymbionts of tsetse flies. *International Journal of Systematic Bacteriology*, **45**, 848–851.

Askoy, S., Chen, X., and Hypsa, V. (1997). Phylogeny and potential transmission routes of midgut-associated endosymbionts of tsetse (Diptera: Glossinidae). *Insect Molecular Biology*, **6**, 183–190.

Aksoy, S., Pourhosseini, A. A., and Chow, A. (1995). Mycetome endosymbionts of tsetse flies constitute a distinct lineage related to Enterobacteriaceae. *Insect Molecular Biology*, **4**, 15–22.

Alexander, R. D. and Borgia, G. (1978). Group selection, altruism, and the levels of organisation of life. *Annual Review of Ecology and Systematics*, **9**, 449–474.

Andreadis, T. G. (1988). *Amblyospora conneticus* sp. nov. (Microsporidae: Amblyosporidae): horizontal transmission in the mosquito *Aedes cantator* and formal description. *Journal of Invertebrate Pathology*, **52**, 90–101.

Andreadis, T. G. (1991). Epizootiology of *Amblyospora conneticus* (Microsporida) in field populations of the saltmarsh mosquito, *Aedes cantator* and the cyclopoid copepod, *Acnthocyclops vernalis*. *Journal of Protozoology*, **37**, 174–182.

Anxolabéhère, D. and Périquet, G. (1983). Système P/M de dysgénésie des hybrides, polymorphisme génétique et évolution des populations de *Drosophila melanogaster*. *Génetique, Sélection, Evolution*, **45**, 31–44.

Armstrong, E. (1977). Transmission of Nosema kingi to offspring of *Drosophila willistoni* during copulation. *Zeitschrift für Parasitenkunde*, **53**, 311–315.

Arunachalam, N. and Curtis, C.F. (1985). Integration of radiation with cytoplasmic incompatibility for genetic control in the *Culex pipiens* complex (Diptera: Culicidae). *Journal of Medical Entomology*, **22**, 648–653.

Ashburner, M. (1989). *Drosophila: a laboratory handbook*. Cold Spring Harbor Laboratory Press, Cold Spring Harbor, NY.

Assem, J. van den and Povel, G. D. E. (1973). Courtship behaviour of some *Muscidifurax* species: a possible example of a recently evolved ethological isolating mechanism. *Netherlands Journal of Zoology*, **23**, 465–487.

Azad, A. F., Sacci, J. B. Jr, Nelson, W. M., Dasch, G. A., Schmidtmann, E. T., and Carl, M. (1992). Genetic characterization and transovarial transmission of a typhus-like rickettsia found in cat fleas. *Proceedings of the National Academy of Sciences of the United States of America*, **89**, 43–46.

Backer, J. S. and Birky, C. W. Jr (1985). The origin of mutant cells: Mechanisms by which *Sacchromyces cerevisiae* produces cells homoplasmic for new mitochondrial mutants. *Current Genetics*, **9**, 627–640.

Baines, S. (1956).The role of the symbiotic bacteria in the nutrition of *Rhodnius prolixus* (Hemiptera). *Journal of Experimental Biology*, **33,** 533–541.

Balas, M., Lee, M. H., and Werren, J. H. (1996). Geographical distribution and fitness effects of the sonkiller bacterium in Nasonia. *Evolutionary Ecology*, **10**, 593–607.

Bandi, C., Damiani, G., Magrassi, L., Grigolo, A., Fani, R., and Sacchi, L. (1994). Flavobacteria as intracellular symbionts in cockroaches. *Proceedings of the Royal Society of London: Series B, Biological Science*s, **25**, 43–48.

Bandi, C., Sironi, M., Damiani, G., Magrassi, L., Nalepa, C. A., Laudani, U., *et al.* (1995). The establishment of intracellular symbiosis in an ancestor of cockroaches and termites. *Proceedings of the Royal Society of London: Series B, Biological Science*s, **259**, 293–299.

Banks, C. J. (1955). An ecological study of Coccinellidae associated with *Aphis fabae* on *V. faba*. *Bulletin of Entomological Research*, **46**, 561–587.

Barr, A. R. (1980). Cytoplasmic incompatibility in natural populations of a mosquito, *Culex pipiens* L. *Nature*, **283**, 71–72.

Barton, N. H. (1979). The dynamics of hybrid zones. *Heredity*, **43**, 341–359.

Battaglia, B. (1963). Deviation from panmixia as a consequence of sex determination in the marine copepod *Tisbe reticulata*. *Genetics Today*, **1**, 926.

Beard, C. B., Mason, P. W., Aksoy, S., Tesh, R. B., and Richards, F. F. (1992). Transformation of an insect symbiont and expression of a foreign gene in the Chagas' disease vector *Rhodnius prolixus*. *American Journal of Tropical Medicine and Hygiene*, **46**, 195–200.

Beard, C. B., O'Neill, S. L., Mason, P., Mandelco, L., Woese, C. R., Tesh, R.B., *et al.* (1993). Genetic transformation and phylogeny of bacterial symbionts from tsetse. *Insect Molecular Biology*, **1**, 123–131.

Bechnel, J. J. and Sweeney, A. W. (1990). *Amblyospora trinus* n. sp. (Microsporida: Amblyosporidae) in the Australia mosquito *Culex halifaxi* (Diptera: Culicidae). *Journal of Protozoology*, **37**, 584–592.

Beckett, E. B., Boothroyd, B., and Macdonald, W. W. (1978). A light and electron microscope study of rickettsia like organisms in the ovaries of mosquitoes of the *A. scutellaris* group. *Annals of Tropical Medicine and Parasitology*, **72**, 277–283.

Beeman, R. W., Friesen, K. S., and Denell, R. E. (1992). Maternal-effect selfish genes in flour beetles. *Science*, **256**, 89–92.

Bell, G. (1982). *The masterpiece of nature. The evolution and genetics of sexuality*. University of California Press, Los Angeles.

Bensaadimerchermek, N., Salvado J. C., Cagnon, C., Karama, S., and Mouches, C. (1996). Characterization of the unlinked 16S rDNA and 23S-5S ribosomal-RNA operon of *Wolbachia pipientis*, a prokaryotic parasite of insect gonads. *Gene*, **165**, 81–86.

Beukeboom, L. W. and Werren, J. H. (1992). Population genetic analysis of a parasitic chromosome: experimental analysis of psr in subdivided populations. *Evolution*, **46**, 1257–1268.

Beukeboom, L. W. Reed, K. M., and Werren, J. H. (1993). Effects of deletions on mitotic stability of the Paternal Sex Ratio (PSR) chromosome from *Nasonia*. *Chromosoma (Berl.)*, **102**, 20–26.

Binnington, K. C. and Hoffmann, A. A. (1989). *Wolbachia*-like organisms and cytoplasmic incompatibility in *Drosophila simulans*. *Journal of Invertebrate Pathology*, **54**, 344–352.

Birky, C. W. (1978). Transmission genetics of mitochondria and chloroplasts. *Annual Review of Genetics*, **12**, 471–512.

Birky, C. W., Fuerst, P., and Maruyama, T. (1989). Organelle gene diversity under migration, mutation, and drift; Equilibrium expectations, approach to equilibrium, effects of heteroplasmic cells, and comparisons to nuclear genes. *Genetics*, **121**, 613–627.

Birova, H. (1970). A contribution to the knowledge of the reproduction of *Trichogramma embryophagum*. *Acta Entomologia Bohemoslovakia*, **67**, 70–82.

Blickenstaff, C. C. (1965). Partial intersterility of Eastern and Western U.S. strains of the alfalfa weevil. *Annals of the Entomological Society of America*, **58**, 523–526.

Boller, E. F., Russs, K., Vallo, V., and Bush, G. L. (1976). Incompatible races of European Cherry Fruit Fly, *Rhagoletis cerasi* (Diptera: Tephritidae), their origin and potential use in biological control. *Entomologia experimentalis et applicata*, **20**, 237–247.

Bourtzis, K., Nirgianaki, A., Markakis, G., and Savakis, C. (1996). *Wolbachia* infection and cytoplasmic incompatibility in *Drosophila* species. *Genetics*, **144**, 1063–1073.

Bourtzis, K., Nirgianaki, A., Onyango, P., and Savakis, C. (1994). A prokaryotic *dnaA* sequence in *Drosophila melanogaster*: *Wolbachia* infection and cytoplasmic incompatibility among laboratory strains. *Insect Molecular Biology*, **3**, 131–142.

Bowen, W. R. and Stern, V. M. (1966). Effect of temperature on the production of males and sexual mosaics in uniparental race of *Trichogramma semifumatum*. *Annals of the Entomological Society of America*, **59**, 823–834.

Boyle, L., O'Neill, S. L., Robertson, H. M., and Karr, T. L. (1993). Interspecific and intraspecific horizontal transfer of *Wolbachia* in *Drosophila*. *Science*, **260**, 1796–1799.

Braig, H. R., Guzman, H., Tesh, R. B., and O'Neill, S. L. (1994). Replacement of the natural *Wolbachia* symbiont of *Drosophila simulans* with a mosquito counterpart. *Nature*, 367, 453455.

Breeuwer, J. A. J. and Jacobs, G. (1996). *Wolbachia*: intracellular manipulators of mite reproduction. *Experimental and Applied Acarology*, **20**, 421–434.

Breeuwer, J. A. J. and Werren, J. H. (1990). Microorganisms associated with chromosome destruction and reproductive isolation between two insect species. *Nature*, **346**, 558–560.

Breeuwer, J. A. and Werren, J. H. (1993). Cytoplasmic incompatibility and bacterial density in *Nasonia vitripennis*. *Genetics*, **135**, 565–574.

Breeuwer, J. A. J., Stouthamer, R., Barns, S. M., Pelletier, D. A., Weisburg, W. G., and Werren, J. H. (1992). Phylogeny of cytoplasmic incompatibility microorganisms in the parasitoid wasp genus *Nasonia* (Hymenoptera: Pteromalidae) based on 16S ribosomal DNA sequences. *Insect Molecular Biology*, **1**, 25–36.

Bregliano, J. C., Gicard, G., Bucheton, A., Pelisson, A., Lavigne, J. M., and L'Héritier, P. (1980). Hybrid dysgenesis in *Drosophila melanogaster*. *Science*, **207**, 606–611.

Bressac, C. and Rousset, F. (1993). The reproductive incompatibility system in *Drosophila simulans*, DAPI-staining analysis of the *Wolbachia* symbionts in sperm cysts. *Journal of Invertebrate Pathology*, **61**, 226–230.

Brimacombe, L. C. (1980). All-female broods in field and laboratory populations of the Egyptian cotton leafworm, *Spodoptera littoralis* (Boisduval) (Lepidoptera: Noctuidae). *Bulletin of Entomological Research*, **70**, 475–481.

Brower, J. H. (1976). Cytoplasmic incompatibility: occurrence in a stored product pest *Ephestia cautella*. *Annals of the Entomological Society of America*, **69**, 1011–1015.

Brown, S. D. (1965) Chromosomal survey of the armored and palm scale insects (Coccoidea: Diaspididae and Phoenicococcidae). *Hilgardia*, **35**, 189–294.

Buchner, P. (1954). Endosymbiosestudien an Schildläusen. I. *Stictococcus sjoestedti*. *Zeitschrift für Morphologie und Ökologie der Tiere*, **43**, 262.

Buchner, P. (1955). Endosymbiosestudien an Schildläusen. II. *Stictococcus diversiseta. Zeitschrift für Morphologie und Ökologie der Tiere*, **43**, 397.

Buchner, P. (1965). *Endosymbiosis of animals with plant microorganisms.* Interscience, New York.

Bull, J. J. (1983). *Evolution of sex determining mechanisms.* Benjamin/Cummings, Menlo Park, California.

Bull, J. J., Molineux, I. J., and Werren, J. H. (1992). Selfish genes. *Science*, **256**, 65.

Bulnheim, H. P. (1978). Interactions between genetic, external and parasitic factors in sex determination of the crustacean amphipod *Gammarus duebeni. Hegolaender Wissenschaftliche Meeresuntersuchungen*, **31**, 1–33.

Bulnheim, H. P. and Vavra, J. (1968). Infection by the microsporidian *Octosporea effeminans* sp. N., and its sex determining influence in the amphipod *Gammarus duebeni. Journal of Parasitology*, **54**, 241–248.

Burgdorfer, W. (1975). Rocky mountain spotted fever. In *Diseases transmitted from animal to man*, (6th edn), (ed. W. Hubber *et al.*). Charles Thomas, Springfield, Illinois.

Busenberg, S. and Cooke, K. (1993). *Vertically transmitted diseases.* Biomathematics, Vol. 23. Springer-Verlag, Berlin.

Cabello, T. and Vargas, P. (1985). Temperature as a factor influencing the form of reproduction of *Trichogramma cordubensis. Zeitschrift für Angewandte Entomologie*, **100**, 434–441.

Caclavanti, A. G. L., Falcao, D. N., and Castro, L. E. (1957). Sex-ratio in *Drosophila prosaltans*—a character due to the interaction between nuclear and cytoplasmic factors. *American Naturalist*, **91**, 327–329.

Callan, E. M. (1940). On the occurrence of males of *Rhodites rosae. Proceedings of the Royal Entomological Society, London (a)*, **15**, 21–26.

Carlson, J., Olson, K., Higgs, S., and Beaty, B. (1995). Molecular genetic manipulation of mosquito vectors. *Annnual Review of Entomology*, **40**, 359–388.

Caspari, E. and Watson, G. S. (1959). On the evolutionary importance of cytoplasmic sterility in mosquitoes. *Evolution*, **13**, 568–570.

Cavalcanti, A. L. G., Falcao, D. N., and Castro, D. L.. (1957). Sex ratio in *Drosophila prosaltansa*—character due to interaction between nuclear genes and cytoplasmic factors. *American Naturalis*t, **91**, 327–329.

Chapela, II. H., Rehner, S. A., Schultz, T. R., and Mueller, U. G. (1994). Evolutionary history of the symbiosis between fungus-growing ants and their fungi. *Science*, **266**, 1691–1694.

Charlesworth, B. (1991). The evolution of sex chromosomes. *Science*, **251**, 1030–1033.

Charniaux-Cotton, H. (1959). Etude comparée du développement post-embryonnaire de l'appareil génital et de la glande androgène chez *Orchestia gammarella* et *Orchestia mediterranea* (Crustacés Amphipodes).

Autodifférenciation ovarienne. *Bulletin de la Société Zoologique de France*, **84**, 105–115.

Charniaux-Cotton, H. and Payen, G. (1985). Sexual differentiation. In *The biology of crustacea*, Vol. 9, (ed. D.E. Bliss), pp. 217–299. Academic Press, New York.

Chen, B. H., Kfir, R., and Chen, C.N. (1992). The thelytokous *Trichogramma chilonis* in Taiwan. *Entomologia experimentalis et applicata*, **65**, 187–194.

Clancy, D. J. and Hoffmann, A. A. (1997). Behavior of *Wolbachia* endosymbionts from *Drosophila simulans* in *D. serrata*, a novel host. *American Naturalist*, **149**, 975–988.

Clarke, C., Sheppard, P. M., and Scali, V. (1975). All female broods in the butterfly *Hypolimnas bolina* (L.). *Proceedings of the Royal Society of London: Series B, Biological Sciences*, **189**, 29–37.

Clausen, C. P. (1940). *Entomophagous insects*. McGraw-Hill, New York.

Clay, K. (1988). Clavipitaceous fungal epiphytes of grasses: coevolution and the change from parasitism to mutualism. In *Coevolution of fungi with plants and animals* (ed. K. A. Pirozynski and D. L. Hawkesworth). Academic Press, London.

Cooper, K. W. (1959). A bilaterally gynandromorphic *Hyodynerus*, and a summary of cytologic origins of such mosaic Hymenoptera. Biology of Eumine wasps VI. *Bulletin of the Florida State Museum of Biological Sciences*, **5**, 25–40.

Cosmides, L. and Tooby, J. (1981). Cytoplasmic inheritance and intragenomic conflict. *Journal of Theoretical Biology*, **89**, 83–129.

Counce, S. J. and Poulson, D. F. (1962). Developmental effects of the sex-ratio agent in embryos of *Drosophila willistoni*. *Journal of Experimental Zoology*, **151**, 17–31.

Crew, F. A. E. (1927). *The genetics of sexuality in animals*. Cambridge University Press, Cambridge.

Crozier, R. H. (1975). Hymenoptera, In *Animal cytogenetics. 3: Insecta* 7, (ed. B. John), pp. 1–95. Borntraeger, Berlin.

Curtis, C. F. (1968). Possible use of translocations to fix desirable genes in insect pest populations. *Nature*, **218**, 368–369.

Curtis, C. F. (1972). Sterility from crosses between sub-species of the tsetse fly Glossina morsitans. *Acta Tropica*, **23**, 250–268.

Curtis, C. F. (1976). Population replacement in *Culex fatigans* by means of cytoplasmic incompatibility. 2. Field cage experiments with overlapping generations. *Bulletin of the World Health Organization*, **53**, 107–119.

Curtis, C. F. (1994). The case for malaria control through the genetic manipulation of its vectors. *Parasitology Today*, **10**, 371–373.

Curtis, C. F. and Adak, T. (1974). Population replacement in *Culex fatigans* by means of cytoplasmic incompatibility 1. Laboratory experiments with non-overlapping generations. *Bulletin of the World Health Organization*, **51**, 249–255.

Curtis, C. F. and Graves, P. M. (1988). Methods for replacement of malaria vector populations. *Journal of Tropical Medicine and Hygiene*, **91**, 43–48.

Curtis, C. F. and Von Borstel, R. C. (1978). Allegations against Indian research refuted. *Nature*, **273**, 96.

Curtis, C. F., Akiyama, J., and Davidson, G. (1976). A genetic sexing system in *Anopheles gambiae* species. *Mosquito News*, **36**, 492–498

Curtis, C. F., Brooks, G. D., Ansari, M. A., Grover, K. K., Krishnamurthy, B. S., Rajagopalan, P. K., *et al.* (1982). A field trial on control of *Culex quiquefasciatus* by release of males of a strain integrating cytoplasmic incompatibility and a translocation. *Entomologia experimentalis et applicata*, **31**, 181–190.

Curtis, C. F., Hill, N., and Kasim, S. H. (1993). Are there effective resistance management strategies for vectors of human disease? *Biological Journal of the Linnean Society*, **48**, 3–18.

Danthanarayana, W. (1983). Population ecology of the light brown apple moth, *Epiphyas postvittana* (Lepidoptera: Tortricidae). *Journal of Animal Ecology*, **52**, 1–33.

DeBach, P. (1969). Uniparental, sibling and semi-species in relation to taxonomy and biological control. *Israel Journal of Entomology*, **4**, 11–27.

de Bary, A. (1879). *Die Erscheinung der Symbiose*. Verlag von Karl J. Trübner, Strassburg.

Dixon, A. F. G. (1970). Factors limiting the effectiveness of the coccinellid beetle *Adalia bipunctata* (L.) as a predator of the sycamore aphid, *Drepanosiphum platanoides* (Schr.). *Journal of Animal Ecology*, **39**, 739–751.

Doncaster, L. (1913). On an inherited tendency to produce purely female families in *Abraxas grossulariata*, and its relation to an abnormal chromosome number. *Journal of Genetics*, **3**, 1–10.

Doncaster, L. (1914). On the relations between chromosomes, sex limited transmission and sex determination in *Abraxas grossulariata*. *Journal of Genetics*, **4**, 1–21.

Doncaster, L. (1922). Further observations on chromosome and sex determination in *Abraxas glossulariata*. *Quarterly Journal of Microbial Science*, **66**, 397–408.

Dunn, A. M., Adams, J., and Smith, J. E. (1993*a*). Transovarial transmission and sex ratio distortion by a microsporidian parasite in a shrimp. *Journal of Invertebrate Pathology*, **61**, 248–252.

Dunn, A. M., Adams, J., and Smith, J. E. (1993*b*). Is intersexuality a cost of environmental sex determination in *Gammarus duebeni*? *Journal of Zoology, London*, **231**, 383–389.

Dunn, A. M., Adams, J., and Smith, J .E. (1994). Intersexuality in the crustacean *Gammarus duebeni. Invertebrate Reproduction and Development*, **25**, 139–142.

Dunn, A. M., Hatcher, M .J., Terry, R. S., and Tofts, C. (1995). Evolutionary ecology of vertically transmitted parasites: transovarial transmission of a microsporidian sex ratio distorter in *Gammarus duebeni. Parasitology*, **111**, S91–S109.

Durvasala, R. V., Gumbs, A., Panackal, A., Kruglov, O., Aksoy, S., Merrifield, R. B., Richards, F. F., and Beard, C. B. (1997). Prevention of insect-borne disease: an approach using transgenic symbiotic bacteria. *Proceedings of the National Academy of Sciences of the United States of America*, **94**, 3274–3278.

Ebbert, M. A. (1991). The interaction phenotype in the *Drosophila willistoni* spiroplasma symbiosis. *Evolution*, **45**, 971–988.

Ebbert, M. A. (1995). Variable effects of crowding on *Drosophila* hosts of male-lethal and non-male-lethal spiroplasmas in laboratory populations. *Heredity*, **74**, 227–240.

Eberhard, W. G. (1980). Evolutionary consequences of intracellular organelle competition. *Quarterly Review of Biology*, **55**, 231–249.

Eskafi, F. M. and Legner, E. F. (1974). Parthenogenetic reproduction in *Hexacola* sp. near *websteri*, a parasite of *Hippelates* eye gnats. *Annals of the Entomological Society of America*, **67**, 76–78.

Ewald, P. W. (1987). Transmission modes and evoluton of parasitism–mutualism continuum. *Annals of the New York Academy of Sciences*, **503**, 295–306.

Farnham, A. W. (1977) Genetics of resistance of houseflies to pyrethroids. *Pesticide Science*, **8**, 631–636

Ferveur, J. F., Strökuhl, K .F., Stocker, R .F., and Greenspan, R. J. (1995). Genetic feminization of brain structures and changed sexual orientation in male *Drosophila. Science*, **267**, 902–905.

Fine, P. E. M. (1975). Vectors and vertical transmission: an epidemiologic perspective. *Annals of the New York Academy of Sciences*, **266**, 173–194.

Fine, P. E. M. (1978). On the dynamics of symbiote-dependent cytoplasmic incompatibility in Culicine mosquitoes. *Journal of Invertebrate Pathology*, **30**, 10–18.

Fisher, R. A. (1930). *The genetical theory of natural selection*. Oxford University Press, Oxford.

Fisher, R. A. (1937). The wave of advance of advantageous genes. *Annals of Eugenics*, **7**, 355–369.

Flanders, S. A. (1944). Uniparentalism in the hymenoptera and its relation to polyploidy. *Science*, **100**, 168–169.

Flanders, S. A. (1945). The bisexuality of uniparental hymenoptera, a function of the environment. *American Naturalist*, **79**, 122–141.

Flanders, S. A. (1950). Races of apomictic parasitic hymenoptera introduced into California. *Journal of Economic Entomology*, **43**, 719–720.

Flanders, S. E. (1965). On the sexuality and sex ratios of hymenopterous populations. *American Naturalist*, **99**, 489–494.

Fleuriet, A. (1982). Factors affecting the frequency of infection by the sigma virus in experimental populations of *Drosophila melanogaster*. *Archives of Virology*, **73**, 121–133.

Fleuriet, A. (1988). Maintenance of a hereditary virus: the Sigma virus in populations of its host, *Drosophila melanogaster*. In *Evolutionary biology* (ed. M. K. Hecht and B. Wallace), pp. 1–30. Plenum, New York.

Fox, G., Wisotzkey, J., and Jurtshuk, P. Jr. (1992). How close is close: 16s rRNA sequence identity may not be sufficient to guarantee species identity. *International Journal of Systematic Bacteriology*, **42**, 166–170.

Frank, S. A. (1989). The evolutionary dynamics of cytoplasmic male sterility. *American Naturalist*, **133**, 345–376.

Frank, S. A. (1994). Kin selection and virulence in the evolution of protocells and parasites. *Proceedings of the Royal Society of London: Series B, Biological Sciences*, **258**, 153–161.

Frank, S. A. (1996). Models of parasite virulence. *Quarterly Review of Biology*, **71**, 37–78.

Frank, S. W. (1983). A hierarchical view of sex-ratio patterns. *Florida Entomologist*, **66**, 42–75.

Freeland, S. J. and McCabe, B. K. (1997). Fitness compensation and the evolution of selfish cytoplasmic elements. *Heredity*, **78**, 391–402.

French, W. L. (1970). Evidence for the segregation of cytoplasmic genes in *Culex pipiens*, an advanced form of animal life. *Genetics*, **64**, S22.

Ganter, P. F. and Hanton, W. K. (1984). A note on the cause of skewed sex ratios in populations of terrestrial isopods in north Carolina. *Crustaceana*, **46**, 154–159.

Geier, P. W. and Briese, D. T. (1980). The light brown apple moth, *Epiphyas postvittana* (Walker). 4. Studies on population dynamics and injuriousness to apples in Australian Capital territory. *Australian Journal of Ecology*, **5**, 63–93.

Geier, P. W., Briese, D. T., and Lewis, T. (1978). The light brown apple moth *Epiphyas postvittana* (Walker). 2. Uneven sex ratios and a condition contributing to them in the field. *Australian Journal of Ecology*, **3**, 467–488.

Ghelelovitch, S. (1952) . Sur le déterminisme génétique de la stérilité dans les croisements entre différentes souches de *Culex autogenicus* Roubaud. *Comptes Rendus de l'Academie des Sciences, Paris, Série III*, **234**, 2386–2388.

Gherna, R. L., Werren, J., Weisburg, W., Cote, R., Woese, C. R., Mandelco, L., and Brenner, D. J. (1991). *Arsenophonus nasoniae* gen.nov., sp.-nov., the causative agent of the son killer trait in the parasitic wasp

Nasonia vitripennis. *International Journal of Systematic Bacteriology*, **41**, 563–565.

Giard, A. (1886). De l'influence de certains parasites Rhizocéphales sur les caractères sexuels de leur hôte. *Comptes Rendus de l'Academie des Sciences, Paris, Série III*, **103**, 84–86.

Ginsburger-Vogel, T. (1975). Temperature-sensitive intersexuality and its determinism in *Orchestia gammarella* (Pallas). In *Intersexuality in the animal kingdom* (ed. R. Reinhboth), pp. 106–120, Springer-Verlag, Berlin.

Ginsburger-Vogel, T. (1991). Intersexuality in *Orchestia mediterranea* Costa, 1853, and *Orchestia aestuarensis* Wildish, 1987 (Amphipoda)—a consequence of hybridation or parasitic infection? *Journal of Crustacean Biology*, **11**, 530–539.

Ginsburger-Vogel, T. and Charniaux-Cotton, H. (1982). Sex determination. In *The biology of Crustacea*, vol. 2, (ed. D.E. Bliss), pp. 257–283, Academic Press, New York.

Ginsburger-Vogel, T. and Desportes, I. (1979). Structure and biology of *Marteilia* sp. in the amphipod *Orchestia gammarellus*. *Marine Fisheries Review*, **41**, 3–7.

Ginsburger-Vogel, T. and Magniette-Mergault, F. (1981*a*). The effects of temperature on sexual differentiation in the temperature sensitive thelygenic-intersexual offspring of *Orchestia gammarellus* (Pallas) (Amphipoda, Crustacea). I—Effects of temperature on pubescent males. *International Journal of Invertebrate Reproduction*, **4**, 39–50.

Ginsburger-Vogel, T. and Magniette-Mergault, F. (1981*b*). The effects of temperature on sexual differentiation in the temperature sensitive thelygenic-intersexual offspring of *Orchestia gammarellus* (Pallas) (Amphipoda, Crustacea). II—Effects of temperature during embryonic and post-embryonic development. *International Journal of Invertebrate Reproduction*, **4**, 51–65.

Giordano, R., O'Neill, S. L., and Robertson, H. (1995). *Wolbachia* infections and the expression of cytoplasmic incompatibility in *Drosophila sechellia* and *D. mauritiana*. *Genetics*, **140**, 130–717.

Godfray, H. C. J. and Grafen, A. (1988). Unmatedness and the evolution of eusociality. *American Naturalist*, **131**, 303–305.

Gooding, R. H. (1990). Postmating barriers to gene flow among species and subspecies of tsetse flies (Diptera: Glossinidae). *Canadian Journal of Zoology*, **68**, 1727–1734.

Gordh, G. and Lacey, L.. (1976). Biological studies of *Plagiomerus diaspidis*, a primary internal parasite of diaspidid scale insects. *Proceedings of the Entomological Society of Washington*, **78**, 132–144.

Gotoh, T., Gomi, K., and Kamoto, T. (1995*a*). Lethal stage of female embryos in crosses between two local populations of the spider mite, *Tetranychus quercivorus* Ehara et Gotoh (Acari: Tetranychidae). *Experimental and Applied Acarology*, **19**, 129–137.

Gotoh, T., Oku, H., Moriya, K., and Odawara, M. (1995*b*). Nucleus-cytoplasm interactions causing reproductive incompatiblity between two populations of *Tetranychus quercivorus* Ehara et Gotoh (Acari: Tetranychidae). *Heredity*, **74,** 405–414.

Gu, H. and Danthanarayana, W. (1990). Age related flight and reproductive performance of the light brown apple moth, *Epiphyas postvittana. Entomologia experimentalis et applicata*, **54**, 109–115.

Hackett, K. J., Lynn, D. E., Williamson, D. L., Ginsberg, A. S., and Whitcomb, R. F. (1985). Cultivation of the *Drosophila spiroplasma. Science*, **232**, 1253–1255.

Haig, D. and Bergstrom, C. T. (1995). Multiple mating, sperm competition and meiotic drive. *Journal of Evolutionary Biology*, **8**, 265–282.

Hale, L. R. and Hoffmann, A .A. (1990). Mitochondrial DNA polymorphism and cytoplasmic incompatability in natural populations of *Drosophila simulans. Evolution*, **44**, 1383–1386.

Hall, D. W. (1985). The distribution of *Amblyospora* (Microspora) sp.-infected oenocytes in adult female *Culex salinarius*: significance for mechanism of transovariol transmission. *Journal of the American Mosquito Control Association*, **1**, 514–415.

Hall, D. W. (1990). Dimorphic development of *Amblyospora* sp. (Microspora: Amblyosporidae) in *Culex salinarius gynandromorphs. Journal of Invertebrate Pathology*, **55**, 291–292.

Hamilton, W. D. (1967). Extraordinary sex ratios. *Science*, **156**, 477–488.

Hamilton, W. D. (1979). Wingless and fighting males in fig wasps and other insects. In *Sexual selection and reproductive competition in insects* (ed. M. S. Blum and N. A. Blum), pp. 167–220. Springer, New York.

Hamilton, W. D. (1993). Inbreeding in Egypt and in this book: a childish perspective. In *The natural history of inbreeding and outbreeding* (ed. N. W. Thornhill), pp. 429–450. University of Chicago Press, Chicago.

Hartl, D. L. and Clark, A. G. (1989). *Principles of population genetics* (2nd edn). Sinauer, Sunderland, MA.

Hatcher, M. J. and Dunn, A. M. (1995). Evolutionary consequences of cytoplasmically inherited feminizing factors. *Transactions of the Royal Society of London: Series B, Biological Sciences*, **348**, 445–456.

Hatcher, M. J., Dunn, A. M., and Tofts, C. (1997). The effect of the embryonic bottleneck on vertical microparasite transmission. In *Computation in cellular and molecular biological systems* (ed. R. Cuthbertson, N. Holcombe, and P. Paton), pp. 339–351. World Scientific, Singapore.

Hazard, E. I. and Anthony, D. W. (1974). A redescription of the genus *Parathelohania* Codreanu 1966 (Microsporida: Protozoa) with a reexamination of previously described species of *Thelohania* Henneguy 1892 and descriptions of two new species of *Parathelohania* from anopholine mosquitoes. *USDA Technical Bulletin*, **1505**, 1–26.

Heath, D. J. and Ratford, J. R. (1990). The inheritance of sex ratio in the isopod, *Sphaeroma rugicauda*. *Heredity*, **64**, 419–425.

Hertig, M. (1936). The rickettsia, *Wolbachia pipientis* (gen. et sp. n.) and associated inclusions of the mosquito *Culex pipiens*. *Parasitology*, **28**, 453–486.

Hertig, M. and Wolbach, S. B. (1924). Studies on rickettsia-like microorganisms in insects. *Journal of Medical Research*, **44**, 329–374.

Heruth, D. P., Pond, F. R., Dilts, J. A., and Quackenbush, R. L. (1994). Characterization of genetic determinants for R body synthesis and assembly in *Caedibacter taeniospiralis* 47 and 116. *Journal of Bacteriology*, **176**, 3559–3567.

Hessler, A. Y., Hessler, R. R., and Sanders, H. L. (1987). Reproductive system of *Hutchinsoniella macracantha*. *Science*, **168**, 1464.

Hoffmann, A. A. (1988). Partial cytoplasmic incompatibility between two Australian populations of *Drosophila melanogaster*. *Entomologia experimentalis et applicata*, **48**, 61–67.

Hoffmann, A. A. and Turelli, M. (1988). Unidirectional incompatibility in *Drosophila simulans*: Inheritance, geographic variation and fitness effects. *Genetics*, **119**, 435–444.

Hoffmann, A. A. Turelli, M., and Simmons, G. M. (1986). Unidirectional incompatibility between populations of *Drosophila simulans*. *Evolution*, **40**, 692–701.

Hoffmann, A. A. Turelli, M., and Harshman, L. G. (1990). Factors affecting the distribution of cytoplasmic incompatibility in *Drosophila simulans*. *Genetics*, **126**, 933–948.

Hoffmann, A. A. Clancy, D. J., and Merton, E. (1994). Cytoplasmic incompatibility in Australian populations of *Drosophila melanogaster*. *Genetics*, **136**, 993–999.

Hoffmann, A. A. Clancy, D. J., and Duncan, J. (1996). Naturally-occurring *Wolbachia* infection in *Drosophila simulans* that does not cause cytoplasmic incompatibility. *Heredity*, **76,** 1–8.

Holden, P. R. Jones, P., and Brookfield, J. F. Y. (1993). Evidence for a *Wolbachia* symbiont in *Drosophila melanogaster*. *Genetical Research*, **62**, 23–29.

Holloway, G. J. (1985). An analysis of inherited factors affecting the sex ratio in the rice weevil, *Sitophilus oryzae* L. *Heredity*, **55**, 145–150.

Horjus, M. and Stouthamer, R. (1995). Does infection with thelytoky-causing *Wolbachia* in the pre-adult and adult life stages influence the adult fecundity of *Trichogramma deion* and *Muscidifurax raptor*? *Proceedings of the section Experimental and Applied Entomology of the Netherlands Entomological Society*, **6**, 35–40.

Howard, H. W. (1942). The genetics of *Armadillidium vulgare* Latr. II. Studies on the inheritance of monogeny and amphogeny. *Journal of Genetics*, **44**, 143–159.

Howard, H. W. (1958). The genetics of *Armadillidium vulgare* Latr. IV. Lines breeding true for amphogeny and thelygeny. *Journal of Genetics*, **56,** 1–10.

Hoy, M. A. (1990). Pesticide resistance in arthropod natural enemies: variability and selection responses. In *Pesticide resistance in arthropods* (ed. R. T. Roush and B. E. Tabashnik), pp. 203–236. Chapman and Hall, New York.

Hsiao, C and Hsiao, T. H. (1985). Rickettsia as the cause of cytoplasmic incompatibility in the alfalfa weevil, *Hypera postica*. *Journal of Invertebrate Pathology*, **45**, 244–246.

Huger, A. M., Skinner, S. W., and Werren, J. H. (1985). Bacterial infections associated with the Son-Killer Trait in the Parasitoid wasp *Nasonia* (= *Mormoniella*) *vitripennis* (Hymenoptera Pteronalidae). *Journal of Invertebrate Pathology*, **46**, 272–280.

Hung, A. C. F., Day, W. H., and Hedlund, R. C. (1988). Genetic variability in arrhenotokous and thelytokous forms of *Mesochorus nigripes*. *Entomophaga*, **33**, 7–15.

Hurst, G. D. D. (1993). Studies of biased sex-ratios in *Adalia bipunctata* L. Unpublished PhD thesis, University of Cambridge.

Hurst, G. D. D. (1997). Wolbachia, cytoplasmic incompatibility, and the evolution of eusociality. *Journal of Theoretical Biology*, **184**, 99–100.

Hurst, G. D. D. and Majerus, M. E. N. (1993). Why do maternally inherited microorganisms kill males? *Heredity*, **71**, 81–95.

Hurst, G. D. D., Majerus, M. E. N., and Walker, L. E. (1992). Cytoplasmic male killing elements in *Adalia bipunctata* (Linnaeus) (Coleoptera: Coccinellidae). *Heredity*, **69**, 84–91.

Hurst, G. D. D., Majerus, M. E. N., and Walker, L. E. (1993). The importance of cytoplasmic male killing elements in natural populations of the two spot ladybird, *Adalia bipunctata* (Linnaeus) (Coleoptera: Coccinellidae). *Biological Journal of the Linnean Society*, **49**, 195–202.

Hurst, G. D. D., Purvis, E. L., Sloggett, J. J., and Majerus, M. E. N. (1994). The effect of infection with male-killing *Rickettsia* on the demography of female *Adalia bipunctata* L. (two spot ladybird). *Heredity*, **73**, 309–316.

Hurst, G. D. D., Sharpe, R. G., Broomfield, A. H., Walker, L. E., Majerus, T. M. O., Zakharov, I. A., and Majerus, M. E. N. (1995). Sexually transmitted disease in a promiscuous insect, *Adalia bipunctata*. *Ecological Entomology*, **20**, 230–236.

Hurst, G. D. D., Hammarton, T. C., Obrycki, J. J., Majerus, T. M. O., Walker, L. E., Bertrand, D., *et al.*. (1996*a*). Male-killing bacteria in a fifth ladybird beetle, *Coleomegilla maculata* (Coleoptera: Coccinellidae). *Heredity*, **77**, 177–185.

Hurst, G. D. D., Sloggett, J. J., and Majerus, M. E. N. (1996*b*). Estimation of the rate of inbreeding in *Adalia bipuncata* L. (Coleoptera: Coccinellidae) from a phenotypic indicator. *European Journal of Entomology*, **93**, 145–150.

Hurst, G. D. D., Majerus, T. M. O., Von Schulenburg, H. G., Ashburner, M., Zakharov, I., and Majerus, M. E. N. (1997*a*) *Spiroplasma* associated with male-killing in Russian populations of *Adalia bipunctata*, in preparation.

Hurst, G. D. D., Hammarton, T. C., Majerus, T. M. O., Bertrand, D., Bandi, C., and Majerus, M. E. N. (1997*b*) The diversity of inherited parasites of insects: the male-killing agent of the ladybird beetle *Coleomegilla maculata* is a member of the Flavobacteria. *Genetical Research*, in press.

Hurst, L. D. (1991). The incidences and evolution of cytoplasmic male-killers. *Proceedings of the Royal Society of London: Series B, Biological Sciences*, **244**, 91–99.

Hurst, L. D. (1992). Intragenomic conflicts as an evolutionary force. *Proceedings of the Royal Society of London: Series B, Biological Sciences*, **248**, 135–140.

Hurst, L. D. (1993). The incidences, mechanisms and evolution of cytoplasmic sex ratio distorters in animals. *Biological Reviews of the Cambridge Philosophical Society*, **68**, 121–193.

Hurst, L. D. and McVean, G. T. (1996). Clade selection, reversible evolution and the persistence of selfish elements: the evolutionary dynamics of cytoplasmic incompatibility. *Proceedings of the Royal Society of London: Series B, Biological Sciences*, **262**, 97–104

Hurst, L. D. and Pomiankowski, A. N. (1991). Causes of sex ratio bias may account for unisexual sterility in hybrids: a new explanation for Haldane's rule and related phenomena. *Genetics*, **128**, 841–858.

Igarashi, S. (1964*a*). A possibility of cytoplasmic inheritance of the male dominancy in *Tigriopus japonicus*. *Science Reports. Tohuku University, Series IV*, **30**, 77–84.

Igarashi, S. (1964*b*). Modification of sex ratio by feeding the ovisacs in *Tigriopus japonicus*. *Science Reports. Tohuku University, Series IV*, **30**, 85–88.

Ikeda, H. (1970). The cytoplasmically-inherited 'sex-ratio' condition in natural and experimental populations of *Drosophila bifasciata*. *Genetics*, **65**, 311–333.

Jaenson, T. G. T. (1986). Sex ratio distortion and reduced lifespan of *Glossina pallidipes* infected with the virus causing salivary gland hyperplasia. *Entomologia experimentalis et applicata*, **41**, 265–271.

James, A. C. and Jaenike, J. (1990). 'Sex ratio' meiotic drive in *Drosophila testacea*. *Genetics*, **126**, 651–656.

Jardak, T., Pintureau, B., and Voegele, J. (1979). Mise en evidence dune nouvelle espèce de Trichogramme. Phénomène dintersexualité, étude enzymatique. *Annales Société Entomologique de France*, **15**, 635–642.

Jenkins, T. M., Babcock, C. S., Geiser, D. M., and Anderson, W. W. (1996). Cytoplasmic incompatibility and mating preference in Colombian *Drosophila pseudoobscura*. *Genetics*, **142**, 189–194.

Johanowicz, D. L. and Hoy, M. A. (1996). *Wolbachia* in a predator-prey system: 16S ribosomal DNA analysis of two Phytoseiids (Acari: Phytoseiidae) and their prey (Acari: Tetranychidae). *Annals of the Entomological Society of America*, **89**, 435–441.

Johnstone, R. A. and Hurst, G. D. D. (1996). Maternally inherited male-killing microorganisms may confound interpretation of mtDNA variation in insects. *Biological Journal of the Linnean Society*, **53**, 453–470.

Jost, E. (1971). Meiosis in the male of *Culex pipiens* and *Aedes albopictus* and fertilization in the *Culex pipiens*-complex. *Canadian Journal of Genetics and Cytology*, **13**, 237–250.

Juchault, P. (1966). Contribution à l'étude de la différenciation mâle chez les crustacés isopodes. Unpublished thesis, Université de Poitiers, France.

Juchault, P. and Legrand, J. J. (1972). Croisements de néo-mâles expérimentaux chez *Armadillidium vulgare* Latr. (Crustacé, Isopode, Oniscoïde). Mise en évidence d'une hétérogamétie femelle. *Comptes Rendus de l'Academie des Sciences, Paris, Série III*, **274**, 1387–1389.

Juchault, P. and Legrand, J. J. (1976*a*). Modification de la sex ratio dans les croisements entre différentes populations du Crustacé Oniscoïde *Armadillidium vulgare* Latr. Notion de déterminisme polygénique et épigénétique du sexe. *Archives de Zoologie Expérimentale et Générale*, **117**, 81–93.

Juchault, P. and Legrand, J. J. (1976*b*). Etude génétique de l'intersexualité des mâles à ouvertures génitales femelles chez l'oniscoïde *Armadillidium vulgare* Latr.: interprétation et modalité de la transmission héréditaire. *Comptes Rendus des Séances de la Société de Biologie*, **170**, 429–433.

Juchault, P. and Legrand, J. J. (1979). Analyse génétique et physiologique de la détermination du sexe dans une population du crustacé isopode oniscoïde *Armadillidium nasatum* (Budde-lund). *Archives de Zoologie Expérimentale et Générale*, **120**, 24–43.

Juchault, P. and Legrand, J. J. (1981*a*). Contribution a l'étude qualitative et quantitative des facteurs controlant le sexe dans les populations du Crustacé Isopode terrestre *Armadillidium vulgare* Latr. II—Populations hébergeant le facteur féminisant F (bactérie intracytoplasmique). *Archives de Zoologie Expérimentale et Générale*, **122**, 65–74.

Juchault, P. and Legrand, J. J. (1981*b*). Contribution à l'étude qualitative et quantitative des facteurs controlant le sexe dans les populations du Crustacé Isopode terrestre *Armadillidium vulgare* Latr. III—Populations n'hébergeant pas le facteur féminisant F. *Archives de Zoologie Expérimentale et Générale,* **122**, 117–131.

Juchault, P. and Legrand, J. J. (1985). Contribution à l'étude du mécanisme de l'état réfractaire à l'hormone androgène chez les *Armadillidium vulgare* Latreille (Crustacé, Isopode, Oniscoïde) hébergeant une bactérie féminisante. *General and Comparative Endocrinology*, **60**, 463–467.

Juchault, P. and Mocquard, J. P. (1989). Effets de l'inoculation d'une bactérie endocellulaire féminisante sur la croissance et la reproduction des femelles du crustacé Oniscoïde *A. vulgare* Latr. Conséquences possibles sur l'évolution des populations. *Crustaceana*, **56**, 83–92.

Juchault, P. and Mocquard, J. P. (1993). Transfer of a parasitic sex factor to the nuclear genome of the host: a hypothesis on the evolution of sex determining mechanisms in the terrestrial Isopod *Armadillidium vulgare* Latr. *Journal of Evolutionary Biology*, **6**, 511–528.

Juchault, P. and Rigaud, T. (1995). Evidence for female heterogamety in two terrestrial crustaceans and the problem of sex chromosomes evolution in isopods. *Heredity*, **75**, 466–471.

Juchault, P., Frelon, M., Bouchon, D., and Rigaud, T. (1994) New evidence for feminizing bacteria in terrestrial isopods: evolutionary implications. *Comptes Rendus de l'Academie des Sciences, Paris, Série III*, **317**, 225–230.

Juchault, P., Legrand, J. J., and Martin, G. (1974) Action interspécifique du facteur épigénétique féminisant responsable de la thélygénie et de l'intersexualité du Crustacé *Armadillidium vulgare* (Isopode Oniscoide). *Annales d'Embryologie et Morphogenese*, **7**, 265–276.

Juchault, P., Legrand, J. J., and Mocquard, J. P. (1980*a*). Contribution à l'étude qualitative et quantitative des facteurs contrtôlant le sexe dans les populations du Crustacé Isopode terrestre *Armadillidium vulgare* Latr. I—La population de Niort (Deux-Sèvres). *Archives de Zoologie Expérimentale et Générale*, **121**, 3–27.

Juchault, P., Martin, G., and Legrand, J. J. (1980*b*). Induction par la température d'une physiologie mâle chez les néo-femelles et les intersexués du Crustacé Oniscoïde *Armadillidium vulgare* Latr. hébergeant un bactéroïde à action féminisante. *International Journal of Invertebrate Reproduction*, **2**, 223–235.

Juchault, P., Rigaud, T., and Mocquard, J. P. (1992). Evolution of sex determining mechanisms in a wild population of *Armadillidium vulgare* Latr. (Crustacea, Isopoda): competition between two feminizing parasitic factors. *Heredity*, **69**, 382–390.

Juchault, P., Rigaud, T., and Mocquard, J. P. (1993). Evolution of sex determination and sex ratio variability in wild populations of *Armadillidium vulgare* Latr. (Crustacea, Isopoda): a case study in conflict resolution. *Acta Oecologica*, **14**, 547–562.

Kambhampati, S., Rai, K. S., and Verleye, D. M. (1992). Frequencies of mitochondrial DNA haplotypes in laboratory cage populations of the mosquito, *Aedes albopictus*. *Genetics*, **132**, 20–59.

Kambhampati, S., Rai, K. S., and Burgun, S. J. (1993). Unidirectional cytoplasmic incompatibility in the mosquito, *Aedes albopictus*. *Evolution*, **47**, 673–677.

Kellen, W. R. and Lindegren, J. E. (1971). Modes of transmission of *Nosema plodiae* Kellen and Lindegren, a pathogen of *Plodia interpuctella* (Hubner). *Journal of Stored Products Research*, **7**, 31–37.

Kellen, W. R., Hoffmann, D. F., and Kwock, R. A. (1981). *Wolbachia* sp (Rickettsiales: Rickettsiaceae) a symbiont of the almond moth *Ephestia cautella*, ultrastructure & influence on host fertility. *Journal of Invertebrate Pathology*, **37**, 273–283.

Kerr, W. E. (1962). Genetics of sex determination. *Annual Review of Entomology*, **7**, 157–176.

King, R. C. (1970). *Ovarian development in* Drosophila melanogaster. Academic Press, New York.

Klunker, R. (1994). The occurrence of puparium parasitoids as natural enemies of house flies. *Applied Parasitology*, **35**, 36–50. [German]

Koch, A. (1967). Insects and their endosymbionts. In: *Symbiosis Vol. II* (ed. S. M. Henry). Academic Press, New York.

Krasfur, E. S., Whitten, C. J., and Novy, J. E. (1987). Screwworm eradication in North and Central America. *Parasitology Today*, **3**, 131–137

Krishnamurthy, B. S. and Laven, H. (1976). Development of cytoplasmically incompatible and integrated (translocated incompatible) strains of *Culex pipiens fatigans* for use in genetic control. *Journal of Genetics*, **62**, 117–129.

LaChance, L. E. (1979). Genetic strategies affecting the success and economy of the sterile insect relese method. In *Genetics in relation to insect management* (ed. M. A. Hoy and J. J. McKelvey Jr.), pp. 8–18. Bellagio, Italy. The Rockefeller Foundation.

Lamb, R. Y. and Willey, R. B. (1989). Parthenogenetic mechanism and its evolutionary potential in the cave cricket *Euhadenoecus insolitus*. *Annals of the Entomological Society of America*, **82**, 101–108.

Lande, R. (1979). Effective deme sizes during long-term evolution estimated from rates of chromosomal rearrangement. *Evolution*, **33**, 234–251.

Laraichi, M. (1978). L'effect de hautes temperatures sur le taux sexuel de *Ooencyrtus fecundus*. *Entomologia experimentalis et applicata*, **23**, 237–242.

Laven, H. (1957). Vererbung durch Kerngene und das Problem der ausserkaryotischen Vererbung bei *Culex pipiens*. II. Ausserkaryotische Vererbung. *Zeitschrift für Vererbungslehre*, **88**, 478–516.

Laven, H. (1959). Speciation by cytoplasmic isolation in the *Culex pipiens*-complex. *Cold Spring Harbor Symposia in Quantitative Biology*, **24**, 166–175.

Laven, H. (1967a). Eradication of *Culex pipiens fatigans* through cytoplasmic incompatibility. *Nature*, **216**, 383–384.

Laven, H. (1967b). Speciation and evolution in *Culex pipiens*. In *Genetics of insect vectors of disease* (ed. J. Wright and R. Pal), pp. 251–275. Elsevier, Amsterdam.

Laven, H. and Aslamkhan, M. (1970). Control of *Culex pipiens pipiens* and *C. p. fatigans* with integrated genetical systems. *Pakistan Journal of Science*, **22**, 303–312.

Lécher, P., Defaye, D., and Noel, P. (1995). Chromosomes and nuclear DNA of Crustacea. *Invertebrate Reproduction and Development*, **27**, 85–114.

Legner, E .F. (1985*a*). Effects of scheduled high temperature on male production in thelytokous *Muscidifurax uniraptor*. *Canadian Entomologist*, **117**, 383–389.

Legner, E. F. (1985*b*). Natural and induced sex ratio changes in populations of thelytokous *Muscidifurax uniraptor*. *Annals of the Entomological Society of America*, **78**, 398–402.

Legner, E. F. (1988). Studies of four thelytokous Puerto Rican isolates of *Muscidifurax uniraptor*. *Entomophaga*, **32**, 269–280.

Legrand, J. J. and Juchault, P. (1969). Le déterminisme de l'intersexualité chez les crustacés isopodes terrestres: corrélation entre intersexualité et monogénie. *Comptes Rendus de l'Academie des Sciences, Paris, Série III*, **268**, 1647–1649.

Legrand, J. J. and Juchault, P. (1970). Modification expérimentale de la proportion des sexes chez les Crustacés Isopodes terrestres: induction de la thélygénie chez *Armadillidium vulgare* (Latr.). *Comptes Rendus de l'Academie des Sciences, Paris, Série III*, **270**, 706–708.

Legrand, J. J. and Juchault, P. (1984) Nouvelles données sur le déterminisme génétique et épigénétique de la monogénie chez le crustacé isopode terrestre *Armadillidium vulgare* Latr. *Génétique, Sélection, Evolution*, **16**, 57–84.

Legrand, J. J., Juchault, P., Mocquard, J. P., and Martin, G. (1980). Polymorphisme d'origine géographique sur la valence des chromosomes sexuels et phénomènes de monogénie résultant du croisement de différentes populations de *Porcellio dilatatus* Brandt (Crustacé, Isopode terrestre). *Reproduction, Nutrition and Development*, **20**, 23–59.

Legrand, J. J., Legrand-Hamelin, E., and Juchault, P. (1987). Sex determination in Crustacea. *Biological Reviews*, **62**, 439–470.

Lehmann, R. and Ephrussi, A. (1994). Germ plasm formation and germ cell determination in *Drosophila*. In: *Germline development*. Ciba Foundation Symposium 182. John Wiley & Sons.

Leigh, E. (1977). How does selection reconcile individual advantage with the good of the group? *Proceedings of the National Academy of Sciences of the United States of America*, **74**, 4542–4546.

Leigh, E. G. (1971). *Adaptation and diversity*. Freeman Cooper & Co., San Francisco.

Leu, S.-J. C. Li, J. K. K., and Hsiao, T.H. (1989). Characterization of *Wolbachia postica*, the cause of reproductive incompatibility among alfalfa weevil strains. *Journal of Invertebrate Pathology*, **54**, 248–259.

Lewis, D. (1941). Male sterility in natural populations of hermaphrodite plants: the equilibrium between females and hermaphrodites to be expected with different types of inheritance. *New Phytologist*, **40**, 56–63.

L'Héritier, P. (1962). Les relations du virus héréditaire de la Drosophile avec son hôte. *Annales de l'Institut Pasteur*, **102**, 511–526.

Lin, M. Y., Treng, Y. J., and Sun, N. (1981). Diamondback moth resistance to several synthetic pyrethroids. *Journal of Economic Entomology*, **74**, 393–396.

Lindquist, D. A., Abusowa, M., and Hall, M. J. R. (1992). The New World screwworm fly in Libya: a review of its introduction and eradication. *Medical and Veterinary Entomology*, **6,** 2–8.

Lipsitch, M., Nowak, M. A., Ebert, D., and May, R. M. (1995). The population dynamics of vertically and horizontally transmitted parasites. *Proceedings of the Royal Society of London: Series B, Biological Sciences*, **260**, 321–327.

Louis, C. and Nigro, L. (1989). Ultrastructural evidence of *Wolbachia* Rickettsiales in *Drosophila simulans* and their relationships with unidirectional cross-incompatibility. *Journal of Invertebrate Pathology*, **54**, 39–44.

Louis, C., Pintureau, B., and Chapelle, L. (1993). Recherches sur l'origine de l'unisexualité: la thermothérapie élimine à la fois rickettsies et parthénogenèse thélytoque chez un Trichogramme (Hym., Trichogrammatidae). *Comptes Rendus de l'Academie des Sciences, Paris, Série III*, **316**, 27–33.

Luck, R. F., Stouthamer, R., and Nunney, L. (1992). Sex determination and sex ratio patterns in parasitic hymenoptera. In *Evolution and diversity of sex ratio in haplodiploid insects and mites* (ed. D. L. Wrench and M. A. Ebbert), pp. 442–476. Chapman & Hall, New York

Macdonald, W. W. (1976). Mosquito genetics in relation to filarial infections. *Symposia. British Society for Parasitology*, **14**, 1–24.

McInnis, D.O., Tam, S., and Miyashita, D. (1994). Population suppression and sterility rates induced by variable sex ratio, sterile insect releases of *Ceratitis capitata* (Diptera:Tephritidae) in Hawaii. *Annals of the Entomological Society of America*, **87**, 231–240.

MacLellan, C. R. (1973). Natural enemies of the light brown apple moth, *Epiphyas postvittana*, in the Australian capital territory. *Canadian Entomologist*, **105**, 681–700.

Magni, G. E. (1954). Thermic cure of cytoplasmic sex-ratio in *Drosophila bifasciata. Caryologia (Suppl.)*, **6**, 1213–1216.

Magnin, M. Pasteur, N., and Raymond, M. (1987). Multiple incompatibilities within populations of *Culex pipiens* L. in southern France. *Genetica*, **74**, 125–130.

Maidak, B. L., Larsen, N., McCaughey, M. J., Overbeek, R., Olsen, G. J., Fogel, K., *et al.* (1994). The Ribosomal Database Project. *Nucleic Acids Research*, **22**, 3485–3487.

Majerus, M. E. N. (1981). All female broods in *Philudoria potatoria* (L.) (Lepidoptera: Lasiocampidae). *Philosophical Transactions of the British Entomology and Natural History Society*, **14**, 87–92.

Majerus, T. M. O., Majerus, M. E. N., Knowles, B., Wheeler, J., Bertrand, D., Zakhorov, I. A., and Hurst, G. D. D. (1997). Variation in male-killing behavior in three populations of the ladybird *Harmonia axyridis*. In preparation.

Malogolowkin, C. (1958). Maternally inherited sex-ratio condition in *Drosophila willistoni* and *Drosophila paulistorum*. *Genetics*, **43**, 274–286.

Malogolowkin, C. (1959). Temperature effects on maternally inherited sex-ratio conditions in *Drosophila willistoni* and *Drosophila equinoxialis*. *American Naturalist*, **93**, 365–368.

Malogolowkin-Cohen, C. and Rodriguez-Pereira, M. A. Q. (1975). Sexual drive of normal and SR flies of *Drosophila paulistorum*. *Evolution*, **29**, 579–580.

Mannicacci, D., Couvet, D., Belhassen, E., Gouyon, P.-H., and Atlan, A. (1996). Founder effects and sex-ratio in the gynodioecious *Thymus vulgaris* L. *Molecular Ecology*, **5**, 63–72.

Margulis, L. (1981). *Symbiosis and cell evolution*. Freeman, New York.

Margulis, L. and Fester, R. (ed.) (1991). *Symbiosis as a source of evolutionary innovation*. The MIT Press

Martin, G., Gruppe, S. G., Laulier, M., Bouchon, D., Rigaud, T., and Juchault, P. (1994). Evidence for *Wolbachia* spp. in the estuarine isopod *Sphaeroma rugicauda* (Crustacea): a likely cytoplasmic sex ratio distorter. *Endocytobiosis and Cell Research*, **10**, 215–225.

Martin, G., Juchault, P., and Legrand, J. J. (1973). Mise en évidence d'un micro-organisme intracytoplasmique symbiote de l'Oniscoïde *Armadillidium vulgare* L., dont la présence accompagne l'intersexualité ou la féminisation totale des mâles génétiques de la lignée thélyghne. *Comptes Rendus de l'Academie des Sciences, Paris, Série III*, **276**, 2313–2316.

Martin, G., Juchault, P., Sorokine, O., and Van Dorsselaer, A. (1990). Purification and characterization of androgenic hormone from the terrestrial isopod *Armadillidium vulgare* Latr. (Crustacea, Oniscidea). *General and Comparative Endocrinology*, **80**, 349–354.

Matsuka, M., Hashi, H., and Okada, I. (1975). Abnormal sex-ratio found in the lady beetle, *Harmonia axyridis* Pallas (Coleoptera: Coccinellidae). *Applied Entomology and Zoology*, **10**, 84–89.

Maudlin, I. (1991). Transmission of African trypanosomiasis interactions among tsetse immune system, symbionts and parasites: a review. *Advances in Disease–Vector Research*, **7**, 117–140.

Maurice, S., Belhassen, E., Couvet, D., and Gouyon, P. H. (1994). Evolution of dioecy: can nuclear-cytoplasmic interactions select for maleness? *Heredity*, **73**, 346–354.

May, R. M. and Anderson, R. M. (1983). Epidemiliology and genetics in the coevolution of parasites and hosts. *Proceedings of the Royal Society of London: Series B, Biological Sciences*, **219**, 281–313.

Maynard Smith, J. (1978). *The evolution of sex*. Cambridge University Press.

Maynard Smith, J. (1991), A darwinian view of symbiosis. In *Symbiosis as a source of evolutionary innovation* (ed. L. Margulis and R. Fester). The MIT Press.

Maynard Smith, J. and Szathmary, E. (1995). *The major transitions in evolution*. W.H. Freeman.

Meek, S. R. (1984). Occurrence of rickettsia like symbionts among species of the *Aedes scutellaris* group Diptera Culicidae. *Annals of Tropical Medicine and Parasitology*, **78**, 377–382.

Meek, S. R. and Macdonald, W. W. (1984). Crossing relationships among seven members of the group of *Aedes scutellaris* (Walker) (Diptera: Culicidae). *Bulletin of Entomological Research*, **74**, 65–78.

Meer, M. M. M. van, Kan, F. J.M. P. van, Breeuwer, J. A.J., and Stouthamer, R. (1995). Identification of symbionts associated with parthenogenesis in *Encarsia formosa* and *Diplolepis rosae*. *Proceedings of the section Experimental and Applied Entomology of the Netherlands Entomological Society*, **6**, 81–86.

Merçot, H., Llorente, B., Jacques, M., Atlan, A., and Montchamp-Moreau, C. (1995). Variability within the Seychelles cytoplasmic incompatibility system in *Drosophila simulans*. *Genetics*, **141**, 1015–1023.

Miller, D. R. and Borden, J. H. (1985). Life history and biology of *Ips latidens* (Leconte) (Coleoptera: Scolytidae). *Canadian Entomology*, **117**, 859–871.

Montchamp-Moreau, C. Ferveur, J.-F., and Jacques, M. (1991). Geographic distribution and inheritance of three cytoplasmic incompatibility types in *Drosophila simulans*. *Genetics*, **129**, 399–407.

Moran, N. A. (1996). Accelerated evolution and Muller's rachet in endosymbiotic bacteria. *Proceedings of the National Academy of Sciences of the United States of America*, **93**, 2873–2878.

Moran, N., and Baumann, P. (1994). Phylogenetics of cytoplasmically inherited microorganisms of arthropods. *Trends in Ecology and Evolution*, **9**, 15–20.

Muller, H. M., Crampton, J. M., della Torre, A., Sinden, R., and Crisanti, A. (1993). Members of a trypsin gene family in *Anopheles gambiae* are induced in the gut by blood meal. *EMBO Journal*, **12**, 2891–2900.

Nauta, M. J. and Hoekstra, R. F. (1993). Evolutionary genetics of Spore Killers. *Genetics*, **135**, 923–930.

Naylor, C., Adams, J., and Greenwood, P.J. (1988). Variation in sex determination in natural populations of a shrimp. *Journal of Evolutionary Biology*, **1**, 355–368.

Ndiaye, M. and Mattei, X. (1993). Endosymbiotic relationship between a rickettsia-like microorganism and the male germ-cells of *Culex tigripes*. *Journal of Submicroscopic Cytology & Pathology*, **25**, 71–77.

Nigro, L. (1991). The effect of heteroplasmy on cytoplasmic incompatibility in transplasmic lines of *Drosophila simulans* showing a complete replacement of the mitochondrial DNA. *Heredity*, **66**, 41–45.

Nigro, L. and Prout, T. (1990). Is there selection on RFLP differences in mitochondrial DNA? *Genetics*, **125**, 551–555.

Noda, H. (1984*a*). Cytoplasmic incompatibility in a rice planthopper. *Journal of Heredity*, **75**, 345–348.

Noda, H. (1984*b*). Cytoplasmic incompatibility in allopatric field populations of the small brown planthopper, *Laodelphax striatellus*, in Japan. *Entomologia experimentalis et applicata*, **35**, 263–267.

Noda, H. (1987). Further studies of cytoplasmic incompatibility in local populations of *Laodelphax striatellus* in Japan (Homoptera: Delphacidae). *Applied Entomology and Zoology*, **22**, 443–448.

Noda, H. and Kodama, K. (1996). Phylogenetic position of yeastlike endosymbionts of anobiid beetles. *Applied and Environmental Microbiology*, **62**, 162–167.

Noda, H., Nakashima, N., and Koizumi, M. (1995). Phylogenetic position of yeast-like symbiotes of rice planthoppers based on partial 18S rDNA sequences. *Insect Biochemistry and Molecular Biology*, **25**, 639–646.

Noor, M. and Coyne, J. (1995). A factor causing distorted sex ratios in *Drosophila simulans*. *Drosophila Information Service*, **76**, 151–152.

Nur, U., Werren, J. H., Eickbush, D. G., Burke, W. D., and Eickbush, T. H. (1988). A selfish B-chromosome that enhances its transmission by eliminating the paternal genome. *Science*, **240**, 512–514.

O'Neill, S. L. (1989). Cytoplasmic symbionts in *Tribolium confusum*. *Journal of Invertebrate Pathology*, **53**, 132–134.

O'Neill, S. L. and Karr, T. L. (1990). Bidirectional incompatibility between conspecific populations of *Drosophila simulans*. *Nature*, **348**, 178–180.

O'Neill, S. L., Giordano, R., Colbert, A. M. E., Karr, T. L., and Robertson, H. M. (1992). 16S rRNA phylogenetic analysis of the bacterial endosymbionts associated with cytoplasmic incompatibility in insects. *Proceedings of the National Academy of Sciences of the United States of America*, **89**, 2699–2702.

O'Neill, S. L., Gooding, R. H., and Aksoy, S. (1993). Phylogenetically distant symbiotic microorganisms reside in *Glossina* midgut and ovary tissues. *Medical and Veterinary Entomology*, **7**, 377–383.

Ohkuma, M. and Kudo, T. (1996). Phylogenetic diversity of the intestinal bacterial community in the termite *Reticulitermes speratus*. *Applied and Environmental Microbiology*, **62**, 461–468.

Ohkuma, M., Noda, S., Horikoshi, K., and Kudo, T. (1995). Phylogeny of symbiotic methanogens in the gut of the termite *Reticulitermes speratus*. *FEMS Microbiology Letters*, **134**, 45–50.

Olson, K. E., Higgs, S., Gaines, P. J., Powers, A. M., Davis, B. S., Kamrud, K. I., *et al.* (1996) Genetically engineered resistance to dengue-2 virus transmission in mosquitoes. *Science*, **272**, 884–886.

Orphanides, G. M. and Gonzalez, D. (1970). Identity of a uniparental race of *Trichogramma pretiosum*. *Annals of the Entomological Society of America*, **63**, 1784–1786

Osawa, N. (1992). Sibling cannibalism in the ladybird beetle *Harmonia axyridis*: fitness consequences for mother and offspring. *Research in Population Ecology*, **34**, 45–55.

Östergren, G. (1945). Parasitic nature of extra fragment chromosomes. *Botaniska Notiser*, **2**, 157–163.

Otieno, L. H., Kokwaro, E. D., Chimtawi, M., and Onyango, P. (1980). Presence of enlarged salivary glands in wild populations of *Glossina pallidipes* in Kenya, with a note on the ultrastructure of the affected organ. *Journal of Invertebrate Pathology*, **36**, 113–118.

Owen, D. F. (1970). Inheritance of sex ratio in the butterfly *Acraea encedon*. *Nature*, **225**, 662–663.

Owen, D. F. and Smith, D. A. S. (1991). All-female broods and mimetic polymorphism in *Acraea encedon* (L.) (Lepidopera: Acraeidae) in Tanzania. *African Journal of Ecology*, **29**, 241–247.

Owen, D. F. (1973). Low mating frequencies in an African butterfly. *Nature*, **244**, 116–117.

Perkins, R. C. L. (1905). Leaf-hoppers and their natural enemies (Mymaridae, Platygasteridae). *Hawaii Sugar Planters Association Experimental Station Bulletin*, **1**, 187–203.

Perrot-Minnot, M-J. Guo, L. R., and Werren, J. H. (1996). Single and double infections with *Wolbachia* in the parasitic wasp *Nasonia vitripennis*: effects on compatibility. *Genetics*, **143**, 961–972.

Petersen, J. J., Watson, D. W., and Pawson, B. M. (1992). Evaluation of field propagation of *Muscidifurax zaraptor* (Hymenoptera: Pteromalidae) for control of flies associated with confined beef cattle. *Journal of Economic Entomology*, **85**, 451–455.

Phillips, E. F. (1903). A review of parthenogenesis. *Proceedings of the American Philosophical Society*, **42**, 275–345.

Pijls, W. A. M., van Steenbergen, H. J., and van Alphen, J. J. M. (1996). Asexuality cured: the relations and differences between sexual and asexual *Apoanagyrus diversicornis*. *Heredity*, **76**, 506–513.

Pinto, J. D., Stouthamer, R., Platner, G. R., and Oatman, E. R. (1991). Variation in reproductive compatibility in *Trichogramma* and its taxonomic significance. *Annals of the Entomological Society of America*, **84**, 37–46.

Pitnick, S., Markow, T. A., and Spicer, G. S. (1995) Delayed male maturity is a cost of producing large sperm in *Drosophila. Proceedings of the National Academy of Sciences of the United States of America*, **93**, 10614–10618.

Pond, F. R., Gibson, I. Lalucat, J., and Quackenbush, R. L. (1989). R Body producing bacteria. *Microbiology Review*, **53**, 25–67.

Preer, J. R., Jr., Preer, L.B., and Jurand, A. (1974). Kappa and other endosymbionts of *Paramecium aurelia. Bacteriology Review*, **38**, 113–163.

Price, G. R. (1970). Selection and covariance. *Nature*, **227**, 520–521.

Price, G. R. (1972). Extension of covariance selection mathematics. *Annals of Human Genetics*, **35**, 485–490.

Priester, T. M. and Georghiou, G. P. (1979). Inheritance of resistance to permethrin in *Culex pipiens quinquefasciatus. Journal of Economic Entomology*, **72**, 124–127.

Prout, T. (1994). Some evolutionary possibilities for a microbe that causes incompatibility in its host. *Evolution*, **48**, 909–911.

Prout, T. and Bungaard, J. (1977). The population genetics of sperm displacement. *Genetics*, **85**, 95–124.

Quezada, J. R., DeBach, P., and Rosen, D. (1973). Biological and taxonomic studies of *Signophora borinquensis*, new species, (Hym: Signiphoridae), a primary parasite of diaspine scales. *Hilgardia*, **41**, 543–604.

Ralph, C. P. (1977). Effect of host plant density on populations of a specialized, seed sucking bug, *Oncopeltus fasciatus. Ecology*, **58**, 799–809.

Raymond, M., Callaghan, A., Fort, P., and Pasteur, N. (1991). Worldwide migration of amplified insecticide resistance genes in mosquitoes. *Nature*, **350**, 151–153.

Reed, K. M. and Werren, J. H. (1995). Induction of paternal genome loss by the paternal-sex-ratio chromosome and cytoplasmic incompatibility bacteria (*Wolbachia*): a comparative study of early embryonic events. *Molecular Reproduction and Development*, **40**, 408–418.

Richardson, P. M. Holmes, W. P., and Saul, G. B. II. (1987). The effect of tetracycline on nonreciprocal cross incompatibility in *Mormoniella* [= *Nasonia*] *vitripennis. Journal of Invertebrate Pathology*, **50**, 176–183.

Rigaud, T. and Juchault, P. (1992). Genetic control of the vertical transmission of a cytoplasmic sex factor in *Armadillidium vulgare* Latr. (Crustacea, Oniscidea). *Heredity*, **68**, 47–52.

Rigaud, T. and Juchault, P. (1993). Conflict between feminizing sex ratio distorters and an autosomal masculinizing gene in the terrestrial isopod *Armadillidium vulgare* Latr. *Genetics*, **133**, 247–252.

Rigaud, T. and Rousset, F. (1996). What generates the diversity of *Wolbachia*–arthropod interactions? *Biodiversity and Conservation*, **5**, 999–1013.

Rigaud, T., Souty-Grosset, C., Raimond, R., Mocquard, J. P., and Juchault, P. (1991*a*). Feminizing endocytobiosis in the terrestrial crustacean

Armadillidium vulgare Latr. (Isopoda): Recent acquisitions. *Endocytobiosis and Cell Research*, **7**, 259–273.

Rigaud, T., Juchault, P., and Mocquard, J. P. (1991*b*) Experimental study of temperature effects on the sex ratio of broods in terrestrial Crustacea *Armadillidium vulgare* Latr. Possible implications in natural populations. *Journal of Evolutionary Biology*, **4**, 603–617.

Rigaud, T., Mocquard, J. P., and Juchault, P. (1992). The spread of parasitic sex factors in populations of *Armadillidium vulgare* Latr (Crustacea, Oniscidea): effects on sex ratio. *Genetics, Selection, Evolution*, **24**, 3–18.

Rigaud, T., Antoine, D., Marcadé, I., and Juchault, P. (1997). The effect of temperature on sex ratio in the isopod *Porcellionides pruinosus*: environmental sex determination or a by-product of cytoplasmic sex determination? *Evolutionary Ecology*, **11**, 205–215.

Roberts, L. W., Rapmund, G., and Cadigan, F. C .J. (1977). Sex ratio in *Rickettsia tsutsugamushi* infected and non-infected colonies of *Leptotrombidium* (Acari: Trombiculidae). *Journal of Medical Entomology*, **14**, 89–92.

Robinson, A. S. and Van Heemert, C. (1982). *Ceratitis capitata*: a suitable case for genetic sexing. *Genetica*, **58**, 229–237.

Roper, D. S. (1979). Distribution of the spider crab, *Leptomithrax longipes* and evidence of bacterially induced feminisation. *New Zealand Journal of Marine and Freshwater Research*, **13**, 303–307.

Rosen, L. (1980). CO_2 sensitivity in mosquitoes infected with sigma, vesicular stomatitis, and other rhabdoviruses. *Science*, **207**, 989–991.

Rœssler, Y. and DeBach, P. (1972). The biosystematic relations between a thelytokous and arrhenotokous form of *Aphytis mytilaspidis*. I. The reproductive relations. *Entomophaga*, **17**, 391–423.

Rœssler, Y. and DeBach, P. (1973). Genetic variability in the thelytokous form of *Aphytis mytilaspidis*. *Hilgardia*, **42**, 149–175.

Rouhbakhsh, D. and Baumann, P. (1995). Characterization of a putative 23S-5S rRNA operon of *Buchnera aphidicola* (endosymbiont of aphids) unlinked to the 16S rRNA-encoding gene. *Gene*, **155**, 107–112.

Rousset, F. and De Stordeur, E. (1994). Properties of *Drosophila simulans* strains experimentally infected by different clones of the bacterium *Wolbachia*. *Heredity*, **72**, 325–331.

Rousset, F. and Solignac, M. (1995). Evolution of single and double *Wolbachia* symbioses during speciation in the *Drosophila simulans* complex. *Proceedings of the National Academy of Sciences of the United States of America*, **92**, 6389–6393.

Rousset, F., Raymond, M., and Kjellberg, F. (1991). Cytoplasmic incompatibilities in the mosquito *Culex pipiens*: How to explain a cytotype polymorphism? *Journal of evolutionary Biology*, **4**, 69–81.

Rousset, F., Bouchon, D., Pintureau, B., Juchault, P., and Solignac, M. (1992*a*). *Wolbachia* endosymbionts responsible for various alterations of sexuality in arthropods. *Proceedings of the Royal Society of London: Series B, Biological Sciences*, **250**, 91–98.

Rousset, F., Vautrin, D., and Solignac, M. (1992b). Molecular identification of *Wolbachia*, the agent of cytoplasmic incompatibility in *Drosophila simulans*, and variability in relation with host mitochondrial types. *Proceedings of the Royal Society of London: Series B, Biological Sciences*, **247**, 163–168.

Rubiliani C., Rubiliani-Durozoi M., and Payen G. (1980). Effets de la sacculine sur les gonades, les glandes androgènes et le système nerveux central des crabes *Carcinus maenas* (L.) et *C. mediterraneus* Czerniavsky. *Bulletin de la Société Zoologique de France*, **105**, 95–100.

Rüttner, F. (1988). *Biogeography and taxonomy of honeybees*. Springer Verlag, Berlin.

Ryan, S. L. and Saul, G. B. II. (1968). Post-fertilization effect of incompatibility factors in *Mormoniella*. *Molecular and General Genetics*, **103**, 29–36.

Ryan, S. L. Saul, G. B., and Conner, G. W. (1985). Aberrant segregation of R-locus genes in male progeny from incompatible crosses in *Mormoniella*. *Journal of Heredity*, **76**, 21–26.

Saffo, M. B. (1991). Symbiogenesis and the evolution of mutualism: Lessons from the Nephromyces-Bacterial endosymbiosis in molgulid tunicates. In *Symbiosis as a source of evolutionary innovation* (ed. L. Margulis and R. Fester). The MIT Press.

Sassaman, C. and Weeks, S. C. (1993). The genetic mechanism of sex determination in the conchostracan shrimp *Eulimnadia texana*. *American Naturalist*, **141**, 314–328.

Saul, G. B. (1961). An analysis of non reciprocal cross incompatability in *Mormoniella vitripennis* (Walker). *Zeitschrift für Vererbungslehre*, **92**, 28–33.

Scali, V. and Masetti, I. (1973). The population structure of *Maniola jurtina* (Lepidoptera: Satyridae). *Journal of Animal Ecology*, **42**, 773–778.

Schilthuizen, M. and Stouthamer, R. (1997). Horizontal transmission of parthenogenesis-inducing microbes in *Trichogramma* wasps. *Proceedings of the Royal Society of London: Series B*, **264**, 361–366.

Schlinger, E. I. and Hall, J. C. (1959). A synopsis of the biologies of three imported parasites of the spotted alfalfa aphid. *Journal of Economic Entomology*, **52**, 154–157.

Shoop, W. L. (1991). Vertical transmission of helminths: hypobiosis and amphiparatenesis. *Parasitology Today*, **7**, 51–54.

Shroyer, D. A. and Rosen, L. (1983). Extrachromosomal inheritance of carbon dioxide sensitivity in the mosquito *Culex quinquefasciatus*. *Genetics*, **104**, 649–659.

Shykoff, J. A. and Schmid-Hempel, P. (1991). Parasites delay worker reproduction in bumble bees: consequences for eusociality. *Behavioural Ecology*, **2**, 242–248.

Silva, I. M. M. S. and Stouthamer, R. (1996). Can the parthenogenesis *Wolbachia* lead to unusual courtship behavior in *Trichogramma*? *Proceedings of the section Experimental and Applied Entomology of the Netherlands Entomological Society*, **7**, 27–31.

Singh, K. R., Curtis, C. F., and Krishnamurthy, B. S. (1976). Partial loss of cytoplasmic incompatibility with age in males of *Culex fatigans*. *Annals of Tropical Medicine and Parasitology*, **70**, 463–466.

Sinkins, S. P., Braig, H. R., and O'Neill, S. L. (1995*a*). *Wolbachia pipientis*: bacterial density and unidirectional cytoplasmic incompatibility between infected populations of *Aedes albopictus*. *Experimental Parasitology*, **81**, 284–291.

Sinkins, S. P., Braig, H. R., and O'Neill, S. L. (1995*b*). *Wolbachia* superinfections and the expression of cytoplasmic incompatibility. *Proceedings of the Royal Society of London: Series B, Biological Sciences*, **261**, 325–330.

Sironi, M., Bandi, C., Sacchi, L., Di Sacco, B., Damiani, G., and Genchi, C. (1995). Molecular evidence for a close relative of the arthropod endosymbiont *Wolbachia* in a filarial worm. *Molecular and Biochemical Parasitology*, **74**, 223–227.

Skinner, S. W. (1982). Maternally inherited sex ratio in the parasitoid wasp *Nasonia vitripennis*. *Science*, **215**, 1133–1134.

Skinner, S. W. (1985). Son-killer: a third extrachromosomal factor affecting sex ratios in the parasitoid wasp *Nasonia vitripennis*. *Genetics*, **109**, 745–754.

Smith, D. (1905). Note on a Gregarine (*Aggregata inachi* n. sp.) which may cause the parasitic castration of its host (*Inachus dorsettensis*). *Mitteilungen. Zoologisches Station Neapel*, **17**, 406–410.

Smith, J. E. and Dunn, A. M. (1991). Transovarial transmission. *Parasitology Today*, **7**, 146–148.

Smith, S. B. (1941). A new form of spruce sawfly identified by means of its cytology and parthenogenesis. *Scientific Agriculture*, **21**, 245–305.

Smith, S. G. (1955). Cytogenetics of obligatory parthenogenesis. *Canadian Entomologist*, **87**, 131–135.

Smith, S. M. (1996). Biological control with *Trichogramma*: advances, successes, and potential of their use. *Annual Review of Entomomlogy*, **41**, 375–406.

Smith-White, S. and Woodhill, A.R. (1954). The nature and significance of non-reciprocal fertility in *Aedes scutellaris* and other mosquitoes. *Proceedings of the Linnean Society of New South Wales*, **79**, 163–176.

Solignac, M. Vautrin, D., and Rousset, F. (1994). Widespread occurrence of the proteobacteria *Wolbachia* and partial cytoplasmic incompatibility in *Drosophila melanogaster*. *Comptes Rendus de l'Academie Sciences Paris, Série III*, **317**, 461–470.

Sorakina, A. P. (1987). Biological and morphological substantiation of the specific distinctness of *Trichogramma telengai* sp. n. *Entomological Review*, **66**, 20–34.

Stanley, J. (1961). Sterile crosses between mutations of *Tribolium confusum*. *Nature*, **191**, 934.

Stevens, L. (1989). Environmental factors affecting reproductive incompatibility in flour beetles, genus *Tribolium*. *Journal of Invertebrate Pathology*, **53**, 78–84.

Stevens, L. and Wade, M. J. (1990). Cytoplasmically inherited reproductive incompatibility in *Tribolium* flour beetles: the rate of spread and effect on population size. *Genetics*, **124**, 367–372.

Stevens, L. and Wicklow, D. T. (1992). Multispecies interactions affect cytoplasmic incompatibility in *Tribolium* flour beetles. *American Naturalist*, **140**, 642–653.

Stille, B. (1985). Population genetics of the parthenogenetic gall wasp *Diplolepis rosae*. *Genetica*, **67**, 145–151.

Stille, B., and Dävring, L. (1980). Meiosis and reproductive stategy in the parthenogenetic gall wasp *Diplolepis rosae*. *Heriditas*, **92**, 353–362.

Stolk, C. and Stouthamer, R. (1995). Influence of a cytoplasmic incompatibility-inducing *Wolbachia* on the fitness of the parasitoid wasp *Nasonia vitripennis*. *Proceedings of the section Experimental and Applied Entomology of the Netherlands Entomological Society*, **7**, 33–37.

Stouthamer, R. (1993). The use of sexual versus asexual wasps in biological control. *Entomophaga*, **38**, 3–6.

Stouthamer, R. and Kazmer, D. J. (1994). Cytogenetics of microbe-associated parthenogenesis and its consequence for gene flow in *Trichogramma* wasps. *Heredity*, **73**, 317–327.

Stouthamer, R. and Luck, R. F. (1991). Transition from bisexual to unisexual cultures in *Encarsia perniciosi*: New data and a reinterpretation. *Annals of the Entomological Society of America*, **84**, 150–157.

Stouthamer, R. and Luck, R. F. (1993). Influence of microbe-associated parthenogenesis on the fecundity of *Trichogramma deion* and *T. pretiosum*. *Entomologia experimentalis et applicata*, **67**, 183–192.

Stouthamer, R. and Werren, J. H. (1993). Microorganisms associated with parthenogenesis in wasps of the genus *Trichogramma*. *Journal of Invertebrate Pathology*, **61**, 6–9.

Stouthamer, R., Luck, R. F., and Hamilton, W. D. (1990*a*). Antibiotics cause parthenogenetic *Trichogramma* to revert to sex. *Proceedings of the National Academy of Sciences of the United States of America* **87**, 2424–2427.

Stouthamer, R., Pinto, J. D., Platner, G. R., and Luck, R. F. (1990*b*). Taxonomic status of thelytokous forms of *Trichogramma* (Hymenoptera: Trichogrammatidae). *Annals of the Entomological Society of America*, **83**, 475–481.

Stouthamer, R., Breeuwer, J. A. J., Luck, R. F., and Werren, J. H. (1993). Molecular identification of microorganisms associated with parthenogenesis. *Nature*, **361**, 66–68.

Stouthamer, R., Luko, S., and Mak, F. (1994). Influence of parthenogenesis *Wolbachia* on host fitness. *Norwegian Journal of Agricultural Sciences*, **16**, 117–122.

Subbarao, S. K., Curtis, C. F., Singh, K. R. P., and Krishnamurthy, B. S. (1974). Variation in cytoplasmic crossing type in a population of *C. p. fatigans* Wied. from the Delhi area. *Journal of Communicable Diseases*, **6**, 80–82.

Subbarao, S. K., Curtis, C. F., Krishnamurthy, B. S., Adak, T., and Chandrahas, R. K. (1977*a*). Selection for partial compatibility with aged and previously mated males in *Culex pipiens fatigans* (Diptera: Culicidae). *Journal of Medical Entomology*, **14**, 82–85.

Subbarao, S. K., Krishnamurthy, B. S., Curtis, C. F., Adak, T., and Chandrahas, R. K. (1977*b*). Segregation of cytoplasmic incompatibility properties in *Culex pipiens fatigans*. *Genetics*, **87**, 381–390.

Suomalainen, E., Saura, A., and Lokki, J. (1987). *Cytology and evolution in parthenogenesis*. CRC Press, Boca Raton, Florida.

Sweeney, A. W., Doggett, S. L., and Gullick, G. (1989). Laboratory experiments on infection rates of *Amblyospora dyxenoides* (Microspora: Ambylosporidae) in the mosquito *Culex annulirostris*. *Journal of Invertebrate Pathology*, **53**, 85–92.

Taviadoraki, P., Benvenuto, E., Trinca, S., De Martinis, D., Cattaneo, A., and Galeffi, P. (1993). Transgenic plants expressing a functional single-chain Fv antibody are specifically protected from virus attack. *Nature*, **366**, 469–472.

Taylor, D. B. and Craig, G. B. (1985). Unidirectional reproductive incompatibility between *Aedes* (Protomacleaya) *brelandi* and *A.* (P.) *hendersoni* (Diptera:Culcidae). *Annals of the Entomological Society of America*, **78**, 769–774.

Taylor, D. R. (1990). Evolutionary consequences of cytoplasmic sex ratio distorters. *Evolutionary Ecology*, **4**, 235–248.

Tesfa-Yohannes, T. M., and Rozeboom, L. E. (1974). Experimental crossing of *Aedes (S.) polynesiensis* Marks. and *A. scutellaris malayensis* Colless (Diptera: Culicidae). *Journal of Medical Entomology*, **11**, 323–331.

Thiriot-Quiévreux, C. and Cuzin-Roudy, J. (1995). Karyological study of the mediterranean krill *Meganyctiphanes norvegica* (Euphausiacea). *Journal of Crustacean Biology*, **15**, 79–85.

Thomson, M. (1927). Studien über die Parthenogenese bei einigen Cocciden und Aleurodiden. *Zeitschrift für Zellforschung und Mikroscopische Anatomie*, **5**, 2–116.

Timberlake, P. H. and Clausen, C. P. (1924). The parasites of *Pseudococcus maritimus* in California. *University of California Publications Technical Bulletin Entomology*, **3**, 223–292.

Trench, R. K. (1991). *Cyanophora paradoxa* Korschikoff and the origins of chloroplasts. In *Symbiosis as a source of evolutionary innovation* (ed. L. Margulis and R. Fester). The MIT Press.

Trpis, M. Perrone, J. B., Ressig, M., and Parker, K. L. (1981). Control of cytoplasmic incompatibility in the *Aedes scutellaris* complex. *Journal of Heredity*, **72**, 313–317.

Tsuyichiya-Omura, S., Sakaguchi, B., Koga, K., and Poulson, D. F. (1985). Morphological features of embryogenesis in *Drosophila melanogaster* infected with a male-killing spiroplasma. *Zoological Science (Tokyo)*, **5**, 373–383.

Turelli, M. (1994). Evolution of incompatibility-inducing microbes and their hosts. *Evolution*, **48**, 1500–1513.

Turelli, M. and Hoffmann, A. A. (1991). Rapid spread of an inherited incompatibility factor in California *Drosophila. Nature*, **353**, 440–442.

Turelli, M., and Hoffmann, A. A. (1995). Cytoplasmic incompatibility in *Drosophila simulans*: dynamics and parameter estimates from natural populations. *Genetics*, **140**, 1319–1338.

Turelli, M., Hoffmann, A. A., and McKechnie, S. W. (1992). Dynamics of cytoplasmic incompatibility and mtDNA variation in natural *Drosophila simulans* populations. *Genetics*, **132**, 713–723.

Uyenoyama, M. K. and Feldman, M. W. (1978). The genetics of sex ratio distortion by cytoplasmic infection under maternal and contagious transmission: an epidemiological study. *Theoretical Population Biology*, **14**, 471–497.

Vandel, A. (1928). La Parthénogenèse géographique: contribution à l'étude biologique et cytologique de la parthénogenèse naturelle. *Bulletin biologique de la France et de la Belgique*, **62**, 164–281.

Vandel, A. (1941). Recherches sur la génétique et la sexualité des isopodes terrestres. VI: Les phénomènes de monogénie chez les oniscoïdes. *Bulletin biologique de la France et de la Belgique*, **75**, 316–363.

Vanderplank, F. L. (1947). Experiments on hybridization of tsetse flies and the possibility of a new method of control. *Transactions of the Royal Entomological Society of London*, **98**, 1–18.

Vazeille, F. M., Ohayon, H., Gounon, P., and Rosen, L. (1992). Unusual morphology of a virus which produces carbon dioxide sensitivity in mosquitoes. *Virus Research*, **24**, 235–247.

Veillet, A. and Graf, F. (1959). Dégénérescence de la glande androgène des crustacés dicapodes parasités par les rhizocéphales. *Bulletin. Société des Sciences Nancy*, **18**, 123–127.

Vetter, R. D. (1991). Symbiosis and the evolution of novel trophic strategies: Thiotrophic organisms at hydrothermal vents. In *Symbiosis as a source of evolutionary innovation*. Margulis, L. and Fester, R. (eds). The MIT Press.

Wade, M. J. (1985). Soft selection, hard selection, kin selection and group selection. *American Naturalist*, **125**, 61–73.

Wade, M. J. and Beeman, R. W. (1994). The population dynamics of maternal-effect selfish genes. *Genetics*, **138**, 1309–1314.

Wade, M. J. and Chang, N. W. (1995). Increased male fertility in *Tribolium confusum* beetles after infection with the intracellular parasite, *Wolbachia. Nature*, **373**, 72–74.

Wade, M. J. and Stevens, L. (1985) Microorganism mediated reproductive isolation in flour beetles (Genus *Tribolium*). *Science*, **227**, 527–528.

Weisburg, W. G., Dobson, M. E., Samuel, J. E., Dasch, G. A., Mallavia, L. P., Baca, O., *et al.* (1989). Phylogenetic diversity of the Rickettsiae. *Journal of Bacteriology*, **171**, 4202–4206.

Werren, J. H. (1987). The coevolution of autosomal and cytoplasmic sex ratio factors. *Journal of Theoretical Biology*, **124**, 317–334.

Werren, J. H. (1991). The paternal sex ratio chromosome of *Nasonia. American Naturalist*, **137**, 392–402.

Werren, J. H. and Beukeboom, L. W. (1993). Population genetics of a parasitic chromosome: theoretical analysis of PSR in subdivided populations. *American Naturalist*, **142**, 224–241.

Werren, J. H. and Jaenike, J. (1995). *Wolbachia* and cytoplasmic incompatibility in mycophagous *Drosophila* and their relatives. *Heredity*, **75**, 320–326.

Werren, J. H., Skinner, S. W., and Charnov, E. L. (1981). Paternal inheritance of a daughterless SR factor. *Nature*, **293**, 467–468.

Werren, J. H., Skinner, S. W., and Huger, A. M. (1986). Male-killing bacteria in a parasitic wasp. *Science*, **231**, 990–992.

Werren, J. H., Hurst, G. D. D., Zhang, W., Breeuwer, J. A. J., Stouthamer, R., and Majerus, M. E. N. (1994). Rickettsial relative associated with male killing in the ladybird beetle (*Adalia bipunctata*). *Journal of Bacteriology*, **176**, 388–394.

Werren, J. H., Zhang, W., and Guo, L. R. (1995*a*). Evolution and phylogeny of *Wolbachia*—reproductive parasites of arthropods. *Proceedings of the Royal Society of London: Series B, Biological Sciences*, **261**, 55–63.

Werren, J. H., Windsor, D., and Guo, L. R. (1995*b*). Distribution of *Wolbachia* among neotropical arthropods. *Proceedings of the Royal Society of London: Series B, Biological Sciences*, **262**, 197–204.

White, M. D. J. (1970). Heterozygosity and genetic polymorphism in parthenogenetic animals. In *Essays in evolution and genetics in honor of Theodosius Dobzhansky* (ed. M. K. Hecht and W. C. Steere). Appleton-Century-Crofts, New York

White, M. J. D. (1973). *Animal cytology and evolution.* Cambridge University Press, Cambridge.

Wildish, D. J. (1971). Adaptative significance of a biased sex ratio in *Orchestia. Nature*, **233**, 54–55.

Wilkins, A. S. (1993). *Genetic analysis of animal development* (2nd edn). Wiley-Liss, New York.

Williams, E. H., Fields, S., and Saul, G. B. (1993). Transfer of incompatibility factors between stocks of *Nasonia (=Mormoniella) vitripennis*. *Journal of Invertebrate Pathology*, **61**, 206–210.

Williams, G. C. (1975). *Adaptation and natural selection. A critique of some current evolutionary thought*. Princeton University Press.

Wilson, F. (1962). Sex determination and gynandromorph production in aberrant and normal strains of *Ooencyrtus submetallicus*. *Australian Journal of Zoology*, **10**, 349–359.

Wilson, F. and Woolcock, L. T. (1960*a*). Environmental determination of sex in a parthenogenetic parasite. *Nature*, **186**, 99–100.

Wilson, F. and Woolcock, L. T. (1960*b*). Temperature determination of sex in a parthenogenetic parasite, *Ooencyrtus submetallicus*. *Australian Journal of Zoology*, **8**, 153–169.

Winger, L., Smith, J. E., Nicholas, J., Carter, E. H., Tirawanchai, N., and Sinden, R. E. (1987). Ookinete antigens of *Plasmodium berghei*: the appearance of a 21kd transmission blocking determinant on the developing ookinete. *Parasite Immunology*, **10**, 193–207.

Winkler, H. (1920). *Verbreitung und Ursache der Parthenogenesis im Pflanzen- und Tierreiche*. Gustav Fisher, Jena.

Woese, C. R. (1987). Bacterial evolution. *Microbiological Reviews*, **51**, 221–271.

Woese, C. R. (1994). There must be a prokaryote somewhere: microbiology's search for itself. *Microbiological Reviews*, **58**, 1–9.

World Health Organization (1995). *Vector control for malaria and other mosquito-borne diseases*. WHO Technical Report Series, 857.

Wratten, S. D. (1973). The effectiveness of the coccinellid beetle, *Adalia bipunctata* (L.) as a predator of the lime aphid, *Eucallipterua tiliae* L. *Journal of Animal Ecology*, **42**, 785–802.

Wratten, S. D. (1976). Searching by *Adalia bipunctata* (L.) (Coleoptera:Coccinellidae) and escape behaviour of its aphid and ciccadellid prey on lime (*Tilia* × *vulgaris* Hayne). *Ecological Entomology*, **1**, 139–142.

Wright, J. D. and Barr, A. R. (1981). *Wolbachia* and the normal and incompatible eggs of *Aedes polynesiensis* (Diptera: Culcidae). *Journal of Invertebrate Pathology*, **38**, 409–418.

Wright, J. D. and Barr, A. R. (1980). The ultrastructure and symbiotic relationships of *Wolbachia* of mosquitoes of the *Aedes scutellaris* group. *Journal of Ultrastructure Research,* **72**, 52–64.

Wright, J. D. and Wang, B. T. (1980). Observations on Wolbachiae in mosquitoes. *Journal of Invertebrate Pathology*, **35**, 200–208.

Wright, J. D., Sjostrand, F. S., Portaro, J. K., and Barr, A. R. (1978). The ultrastructure of the rickettsia-like microorganisms *Wolbachia pipientis* and associated virus-like bodies in the mosquito *Culex pipiens*. *Journal of Ultrastructure Research*, **63**, 79–85.

Yen, J. H. and Barr, A. R. (1971). New hypothesis of the cause of cytoplasmic incompatibility in *Culex pipiens*. *Nature*, **232**, 657–658

Yen, J. H. and Barr, A. R. (1973). The etiological agent of cytoplasmic incompatibility in *Culex pipiens*. *Journal of Invertebrate Pathology*, **22**, 242–250.

Yen, J. H. and Barr, A. R. (1974). Incompatibility in *Culex pipiens*. In *The use of genetics in insect control* (ed. R. Pal and M. J. Whitten), pp. 97–118. Elsevier-North Holland, Amsterdam.

Zakharov, I. A., Hurst, G. D. D., Majerus, M. E. N., and Chersheva, N. (1996). Male-killing in the St Petersburg population of *Adalia bipunctata* is not caused by a Rickettsia. *Russian Journal of Genetics*, in press.

Zchori-Fein, E., Rousch, R. T., and Hunter, M. S. (1992). Male production induced by antibiotic treatment in *Encarsia formosa*. *Experentia*, **48**, 102–105.

Zchori-Fein, E., Rosen, D., and Roush, R. T. (1994). Microorganisms associated with thelytoky in *Aphytis lignanensis*. *International Journal of Insect Morphology and Embryology*, **23**, 169–172.

Zchori-Fein, E., Faktor, O., Zeidan, M., Gottlieb, Y., Czosnek, H., and Rosen, D (1995) Parthenogenesis-inducing microorganisms in *Aphytis*. *Insect Molecular Biolology*, **4**, 173–178.

Zheng, L., Collins, F. H., Kumar, V., and Kafatos, F. C. (1993). A detailed genetic map for the X chromosome of the malaria vector, *Anopheles gambiae*. *Science*, **261**, 605–608.

Index